Nutritional Bioavailability of Zinc

George E. Inglett, EDITOR

U.S. Department of Agriculture
Northern Regional Research Center

Based on a symposium sponsored

by the ACS Division

of Agricultural and Food Chemistry

at the 183rd Meeting

of the American Chemical Society,

Las Vegas, Nevada,

March 28–April 2, 1982

ACS SYMPOSIUM SERIES **210**

AMERICAN CHEMICAL SOCIETY

WASHINGTON, D.C. 1983

Library of Congress Cataloging in Publication Data

Nutritional bioavailability of zinc.
  (ACS symposium series; 210)

  "Based on a symposium sponsored by the ACS
Division of Agricultural and Food Chemistry at the
183rd meeting of the American Chemical Society,
Las Vegas, Nevada, March 28–April 2, 1982."

  Includes index.

  1. Zinc deficiency states—Congresses. 2. Zinc in
the body—Congresses.
  I. Inglett, G. E., 1928– II. American Chemical
Society. Division of Agricultural and Food Chemistry. III. Series.
[DNLM: 1. Biological availability—Congresses.
2. Zinc—Analysis—Congresses. 3. Nutrition—Congresses. 4. Zinc—Metabolism—Congresses. QV 298
N976 1982]

RC627.Z5N87 1982   616.3'96   82-22706
ISBN O-8412-0760-7   ACSMC8   1–280
                                          1983

ACS Symposium Series

M. Joan Comstock, *Series Editor*

FOREWORD

The ACS SYMPOSIUM SERIES was founded in 1974 to provide a medium for publishing symposia quickly in book form. The format of the Series parallels that of the continuing ADVANCES IN CHEMISTRY SERIES except that in order to save time the papers are not typeset but are reproduced as they are submitted by the authors in camera-ready form. Papers are reviewed under the supervision of the Editors with the assistance of the Series Advisory Board and are selected to maintain the integrity of the symposia; however, verbatim reproductions of previously published papers are not accepted. Both reviews and reports of research are acceptable since symposia may embrace both types of presentation.

CONTENTS

Preface vii

1. **Experimental Zinc Deficiency in Humans: An Overview of Original Studies** 1
   Ananda S. Prasad

2. **Trends in Levels of Zinc in the U.S. Food Supply, 1909–1981** 15
   Susan O. Welsh and Ruth M. Marston

3. **Assessment of the Bioavailability of Dietary Zinc in Humans Using the Stable Isotopes ^{70}Zn and ^{67}Zn** 31
   Judith R. Turnlund and Janet C. King

4. **Stable Isotope Approaches for Measurement of Dietary Zinc Availability in Humans** 41
   Morteza Janghorbani, Nawfal W. Istfan, and Vernon R. Young

5. **Zinc Absorption in Humans: Effects of Age, Sex, and Food** 61
   R. L. Aamodt, W. F. Rumble, and Robert I. Henkin

6. **A Redefinition of Zinc Deficiency** 83
   Robert I. Henkin and R. L. Aamodt

7. **Utilization of Zinc by Humans** 107
   S. J. Ritchey and L. Janette Taper

8. **Zinc Bioavailability from Vegetarian Diets: Influence of Dietary Fiber, Ascorbic Acid, and Past Dietary Practices** 115
   C. Kies, E. Young, and L. McEndree

9. **Effect of Fiber and Oxalic Acid on Zinc Balance of Adult Humans** 127
   June L. Kelsay

10. **The Role of Phytate in Zinc Bioavailability and Homeostasis** 145
    D. Oberleas

11. **Dietary Phytate/Zinc Molar Ratio and Zinc Balance in Humans** . . 159
    Eugene R. Morris and Rex Ellis

12. **Zinc Bioavailability from Processed Soybean Products** 173
    John W. Erdman, Jr., Richard M. Forbes, and Hiromichi Kondo

13. **Zinc Bioavailability from Cereal-Based Foods** 185
    G. S. Ranhotra and J. A. Gelroth

14. **Zinc Bioavailability in Infant Formulas and Cereals** 197
    B. G. Shah and B. Belonje

15. **Zinc Absorption from Composite Meals** 211
    Wenche Frølich and Brittmarie Sandstrøm

16. **Zinc Balances in Humans During Different Intakes of Calcium and Phosphorus** 223
    Herta Spencer, Lois Kramer, and Dace Osis

17. **Zinc Transport by Isolated, Vascularly Perfused Rat Intestine and Intestinal Brush Border Vesicles** 233
    Michael P. Menard, Paul Oestreicher, and Robert J. Cousins

18. **Competitive Mineral–Mineral Interaction in the Intestine: Implications for Zinc Absorption in Humans** 247
    Noel W. Solomons

Index 273

PREFACE

ALTHOUGH ZINC WAS KNOWN as a required mineral nutrient for the diets of animals, zinc deficiency in humans' diets was not recognized until the early 1960s. Individuals consuming an amount of dietary zinc exceeding the usual designated requirement still may show signs of nutritional zinc deficiency. Thus, the adequacy of zinc in humans' diets must be evaluated based on the bioavailability of dietary zinc.

This book is based on the symposium that was designed to assess the current perspective and future direction of research on the nutritional bioavailability of zinc. Inhibitors suspected of interfering with the absorption of zinc are some factors that influence bioavailability. Phytates, dietary fibers, proteins, nonenzymatic browning products, and certain micronutrients are among these substances. These inhibitors are covered in various chapters.

The sites of zinc absorption in the mammalian gastrointestinal tract are largely unknown. However, zinc absorption appears to be facilitated by a low molecular weight, zinc-binding ligand. Citric acid and picolinic acid are two such substances with supporting data for their ligand role. Intercellular events involve a zinc flux from mucosa-to-serosa and return. Thionein contributes to the regulation of the entry of dietary zinc from the mucosal cell into the body. These and other biochemical and metabolic aspects of zinc are explained.

Methods for studying zinc bioavailability in humans include metabolic balance studies, radioisotopic techniques, stable isotope techniques, circulating zinc response, and perfusion techniques. These methods of bioavailability, along with other zinc-related studies, also are covered in detail.

I wish to acknowledge the contributions of the authors and reviewers who made this volume possible. My thanks and gratitude are extended to the ACS Division of Agricultural and Food Chemistry and to its Nutrition Chemistry Subdivision for their support.

GEORGE E. INGLETT

U.S. Department of Agriculture
Northern Regional Research Center
Peoria, IL 61604

October 1, 1982

Experimental Zinc Deficiency in Humans

An Overview of Original Studies

ANANDA S. PRASAD

Wayne State University School of Medicine, Harper-Grace Hospitals, Department of Internal Medicine, Detroit, MI 48201, and Veterans Administration Medical Center, Allen Park, MI 48101

During the past two decades, essentiality of zinc for man has been established. Deficiency of zinc in man due to nutritional factors and several disseased states, has been recognized. A marginal deficiency of zinc appears to be prevalent in many segments of population in developed countries and more severe deficiencies are widespread in many parts of the world. In our experimental human model, a marginal deficiency of zinc was induced by dietary means. Loss of body weight (less than 10% in six months on zinc restricted diet), testicular hypofunction, hyperammonemia and a decrease in plasma, urinary and neutrophil zinc concentration were observed. Changes in zinc dependent enzymes such as deoxythymidine kinase in newly synthesized connective tissue and plasma alkaline phosphatase were also observed as a result of zinc restriction and repletion in our model.

Although the role of zinc in human subjects has been now defined and its deficiency recognized in several clinical conditions, these examples are not representative of a pure zinc deficient state in man. It was, therefore, considered desirable to develop a human model which would allow a study of the effects of a mild zinc deficient state in man. Recently such a model has been established successfully in human volunteers with the use of a semi-purified diet based on texturized soy protein.

Long-standing nutritional zinc deficiency has been reported to cause primary hypogonadism in human subjects (1, 2). Deficiency of zinc occurring in association with certain diseases has also been reported to affect adversely testicular function (3). In experimental animals, zinc-restricted diet is known to produce primary hypogonadism (4).

In both animals and human subjects, testicular hypofunction

0097-6156/83/0210-0001$06.00/0

due to zinc deficiency was characterized by decreased function of the Leydig cells and oligospermia (3). A decrease in serum androgens, an increase in serum gonadotropins, and arrest of spermatogenesis were observed in testicular biopsies obtained from some patients who were zinc-deficient. Also, zinc supplementation has been used to increase plasma testosterone level and sperm count in infertile men (5).

In previous reports, a deficiency of zinc was believed to cause male hypogonadism, but because of the complicated nature of the clinical problems in all such cases, other factors responsible for hypogonadism, nutritional or otherwise, could not be ruled out. However, in all such cases reported so far, moderately severe deficiency of zinc was present for several years. In this study we have shown that even a mild deficiency of zinc can adversely influence testicular function in adult subjects, which was reversible with zinc supplementation.

In this paper, I will summarize our experience with this model.

Methods and Results

Four male volunteers participated in the first experiment. The first pair of two subjects (Patient 1, Patient 2) were 57 and 55-year-old white men, respectively. Patient 1 had degenerative osteoarthritis and allergic rhinitis. Results of physical examination, routine blood tests, and zinc concentration of plasma erythrocytes and hair were within normal limits before the study. Patient 2 was diagnosed as having mild diabetes mellitus and mild hypertension. At the time the patients entered the study, there were no abnormal physical findings, and zinc status was within normal limits.

The second pair of subjects were Patient 3, a 56-year-old black man who was being followed up for essential hypertension and gouty arthritis, and Patient 4, a 65-year-old white man who had chronic sinusitis and mild diabetes insipidus. At the time of the study Patient 3 and 4 were asymptomatic, with no abnormal physical or laboratory findings. Zinc status as assessed by zinc concentration in plasma, erythrocytes, and hair was within normal limits in both subjects.

After a thorough physical examination, routine laboratory tests were done, including routine hematologic tests, serum electrolytes, blood urea nitrogen, serum creatinine, and fasting blood sugar. Serum levels of calcium, inorganic phosphate, total cholesterol, triglyceride, total protein, uric acid, total bilirubin, serum glutamic-oxalacetic transaminase, vitamin A, and carotene were also measured. These routine blood tests were repeated once a month. Routine chest roentgenograms and urinalysis were carried out in the beginning and periodically as needed.

The subjects were kept on the metabolic ward, and two physicians followed them up regularly. Psychologic testing was done

by a clinical psychologist thrice, initially, once at the end of
the restricted zinc intake period and again after zinc supplementation.

A semi-purified diet based on texturized soy protein (pur-
chased from General Mills Company, Minneapolis, Minnesota
(Bontrae Products) and Worthington Foods Company, Division of
Miles Laboratory, Elkhart, Indiana) was developed for this study.
Soy protein isolate was used as soy flour in the baked goods
(purchased from General Biochemicals, Teklad Mills, Chagrin
Falls, Ohio). The texturized soy meals used were hamburger
granules, chicken slices, turkey slices, and chicken chunks.
The texturized soy protein and soy protein isolate were washed
twice with ethylenediaminetetraacetate, then rinsed three times
with deionized water, boiled for 30 min. and kept frozen until
ready to be used. Most recipes used in this study were adapted
from Soy Protein Recipe Ideas, published by Institution Volume
Feeding Management Magazine, Chicago, Illinois.

The foods were cooked in large quantities and stored in a
freezer for 1 to 3 months. As needed, the food items were de-
frosted in the refrigerator, heated, weighed, and then served
to the volunteers. Every 4 weeks for 7 days, breakfast, lunch,
dinner, and snacks served each day were homogenized in a blender.
Aliquots of the composite homogenized meals were weighed, frozen,
lyophilized, and then analyzed for fat, nitrogen, and zinc content of the meals. For the first two subjects (Patient 1 and 2),
the diet supplied 1865 kcal, 53 g of protein, and all essential
vitamins and minerals except zinc, according to recommended dietary allowances. The second set of patients (Patients 3 and 4)
received a vitamin tablet (Poly-Vi-Sol, Mead Johnson Laboratories, Evansville, Indiana) and a mineral mixture. The second pair
of subjects also received protein supplement (Stuart Amino Acids
Powder, Stuart Pharmacy, Wilmington, Delaware). The daily intake
of calories was 2352 kcal and protein, 58 g.

The first pair of subjects (Patients 1 and 2) received hospital diet for 2 weeks; then they received the experimental diet
with 10 mg of supplemental zinc (as zinc acetate) daily orally
for 6 weeks. After this, they were given only experimental diet
(daily zinc intake of 2.7 mg) for 24 weeks. At the end of this
phase, while continuing the experimental diet, the two subjects
received 30 mg of zinc supplement (as zinc acetate) daily orally
for 12 weeks. Finally, these subjects were maintained on hospital diet with total daily intake of 10 mg zinc plus 30 mg of oral
zinc supplement (as zinc acetate) for 8 weeks. The hospital diet
provided the same amount of calories and protein as the experimental diet. Thus, these two subjects were observed for a total
period of 52 weeks.

The second two subjects (Patients 3 and 4) received hospital
diet (10 mg of zinc intake daily) for 3 weeks, followed by the
experimental diet with 30 mg of oral zinc supplement (as zinc
acetate) for 5 weeks. After this, they were given only experi-

mental diet (3.5 mg of zinc intake daily) for 50 weeks. The
repletion phase was begun with the administration of 30 mg of
zinc (as zinc acetate) orally while the same experimental diet
was maintained and continued for a total period of 8 weeks, at
the end of which the hospital diet replaced the experimental
diet. The calorie and protein intake from the hospital and the
experimental diets were the same. Oral zinc supplement (30 mg
as zinc acetate) was continued along with the hospital diet for
a period of 8 weeks. Altogether these two subjects were observed
for 64 weeks.

Deionized water was given for drinking purposes throughout
the study period. Care was taken to ensure that the diet provided the same number of calories and protein quantity throughout the study period, including the period in which the subjects
received the hospital diet. Whenever appropriate, extra trays
of the hospital diet (composite of 7 days) were homogenized and
analyzed for fat, protein, and zinc content.

The patients were weighed three times a week, and skinfold
measurements were taken periodically of the abdomen, triceps,
subscapular areas, using skinfold calipers according to the
procedure outlined by the manufacturer (Cambridge Scientific
Industries, Inc., Cambridge, Maryland).

Urine and fecal samples were collected for 7 consecutive
days every 4 weeks. Fecal samples were analyzed for nitrogen,
fat, and zinc content. Fecal samples were weighed, lyophilized,
digested with nitric acid, diluted to volume with deionized
water, and analyzed by the atomic absorption spectrophotometer,
model 303 or 306 (Perkin Elmer, Norwalk, Connecticut). Food
samples were weighed, wet digested with nitric acid, and diluted
to volume with deionized water, then analyzed for zinc level (6).
Nitrogen levels of dried samples (food or feces) were determined
by Kjeldahl procedure (7). The fat content of dried samples
(food or feces) was ascertained by ether extraction of the
lipids (8).

Blood samples were drawn every other week, using plastic
syringes and tubes. The whole blood was centrifuged and plasma
pipetted into plastic tubes, then precipitated with 10% trichlo-
roacetic acid, diluted 1:4 with deionized water, and analyzed
for zinc by the atomic absorption spectrophotometer, model 303
or 306. The erythrocytes were washed three times with normal
saline. After the last centrifugation 1 ml of packed erythrocytes were pipetted into plastic tubes, and 2 ml deionized water
was added. Hemoglobin level was measured using cyanide technique
on Beckman DK-2 spectrophotometer (Fullerton, California).
Erythrocytes were digested with nitric acid and diluted to
volume with deionized water. Zinc was then assayed by atomic
absorption spectrophotometer and values expressed in termes of
micrograms of zinc per gram of hemoglobin. The leukocytes were
separated by a technique reported by Rothstein, Bishop, and
Ashenbrucker (9), and zinc content was measured by atomic absorption spectrophotometer.

For plasma ammonia levels, whole blood was centrifuged and plasma pipetted into ammonia-free tubes. The plasma was then frozen immediately and kept frozen until ammonia was measured by an autoanalyzer. Urea nitrogen in the plasma and creatinine excretion in the urine were also measured by the autoanalyzer. Lactic dehydrogenase activity in the plasma was measured by a method published by Bergmeyer, Bernt, and Hess (10). Ribonuclease activity was ascertained by a modified Sekine method (11, 12). The activity of alkaline phosphatase in the plasma was measured by a colorimetric procedure.

In each of the first two subjects (Patients 1 and 2), after local anesthesia, one polyvinyl sponge measuring approximately 4 cm x 3 cm x 3 mm was implanted subcutaneously on the lateral aspect of the chest. This was done in order to obtain newly synthesizing collagen connective tissue for assay of deoxythymidine kinase activity. Twenty-one days after implantation the sponges were isolated by blunt dissection , and the capsules surrounding the sponges were collected for study. Total protein, total collagen, ribonucleic acid (RNA) to deoxyribonucleic acid (DNA) ratio, and the activity of deoxythymidine kinase were assayed by techniques reported previously (13). Sponge implantation was done twice, once at the end of the zinc restriction period and again after zinc supplementation for 12 weeks, while the subjects received the same experimental diet otherwise.

Clinical and psychologic evaluations during the study period remained essentially unchanged. The first pair of subjects (Patients 1 and 2), on 2.7 mg daily zinc intake, complained of mild roughening of skin and lethargy; but these were not observed in the second pair (Patients 3 and 4), on 3.5 mg of daily zinc intake.

Results of routine laboratory tests remained essentially the same in all four subjects throughout the study except for blood urea nitrogen, which decreased significantly soon after the subjects started receiving the experimental diet. This change occurred even before the zinc intake was restricted, suggesting that the decrease in the blood urea nitrogen was due to the change from animal protein to cereal protein in the diet.

In the first two subjects (Patients 1 and 2), who received 2.7 mg of zinc daily, the weight loss was more pronounced in comparison with the second pair of subjects (Patients 3 and 4), who received 3.5 mg of zinc daily. After repletion with zinc, the weight stabilized in three out of four subjects; and in one subject (Patient 1), although the weight loss continued during the zinc-supplemented period, the rate of weight loss as determined by the slope of the curve was decreased.

The changes in body weight correlated highly with the subscapular thickness in the two subjects in whom these data were obtained (r = 0.839, P < 0.001; r = 0.938, P < 0.001). Further calculation (14), showed that the weight loss could be

accounted for as follows: 50% fat, 30% water, and 20% other
(approximately). In two subjects the intake of fat was 108.3
g/day, and fecal fat (mean ± SD) was 5.6 ± 1.33 g/day in one subject (Patient 4) and in the other 4.11 ± 0.94 g/day throughout
the study period. Thus no effect of zinc depletion-repletion was
observed on fat balance.

The nitrogen excretion in the feces, urine, and the balance
throughout the experimental study period in the second pair of
subjects (Patients 3 and 4) showed no remarkable changes. The
balance data were apparent balances, inasmuch as the nitrogen
excretion in the sweat was not considered in calculating these
balances.

Urinary excretion of zinc decreased in three out of four
subjects as a result of zinc restriction. In one case (Patient
4) there was no decrease in urinary zinc excretion. This was
due to the diuretic therapy (hydrochlorothiazide) that he received for mild hypertension during the study.

In the first two subjects, during the zinc restriction phase
(2.7 mg zinc daily intake), the apparent negative balance for
zinc ranged from 1 to 4 mg/day, whereas in the second group of
subjects the apparent negative balance for zinc was 1 to 2 mg/
day. After supplementation with 30 mg of zinc, the positive zinc
balance ranged from 11 to 22 mg/day in the first pair of subjects, suggesting a retention of approximately 33% to 70% of
zinc intake. In the second pair of subjects (Patients 3 and 4),
during the base-line period when the daily zinc intake was 33.5
mg, the positive balance for zinc was 3 to 4 mg daily. On the
other hand, these subjects, on the same level of zinc intake
(33.5 mg daily) after zinc depletion phase, showed a positive
zinc balance of 14 to 16 mg daily.

The plasma zinc level decreased significantly in all 4 subjects as a result of zinc restriction and increased after supplementation with zinc. The changes were more marked from patients
on 2.7 mg daily zinc intake compared with those on 3.5 mg daily
zinc intake. The erythrocyte zinc level decreased significantly
in the first group of two subjects (Patients 1 and 2) although
the decrease was not evident until 12 weeks on the restricted
zinc intake. In the second group of subjects, although the
erythrocyte zinc level did not decrease significantly during the
zinc-restricted period, it showed a marked increase after zinc
supplementation. Leukocyte zinc decreased significantly due to
zinc restriction in the second group of subjects in whom this
variable was measured.

Plasma alkaline phosphatase was monitored carefully in the
second group of subjects. In both cases, the activity slowly
declined as a result of zinc restriction, and after supplementation with zinc, the activity nearly doubled in 8 weeks. In all
four subjects, the activity of plasma ribonuclease was almost
twice as great during the zinc-restricted period as in the zinc-
supplemented phase. Plasma lactic dehydrogenase activity de-

creased with zinc restriction and increased during supplementation phases in two subjects. Plasma ammonia levels were higher during zinc restriction and decreased after zinc supplementation in two subjects (Patients 3 and 4).

In the sponge connective tissue of the first pair of subjects, total protein and total collagen increased significantly during the zinc supplementation phase in comparison with the zinc restriction phase (Patient 1, restriction phase: total protein, 108 mg; total collagen, 29.3 mg; versus supplementation phase: total protein, 226 mg; and total collagen, 58.4 mg; Patient 2, restriction phase: total protein, 94 mg; total collagen 0.013 mg; versus supplementation phase: total protein, 240 mg; total collagen, 121.2 mg). The RNA-DNA ratio in the sponge connective tissue also increased after zinc supplementation (Patient 1, restriction phase: 0.69 versus supplementation phase: 0.82; Patient 2, restriction phase: 0.71 versus supplemental phase: 0.95). There was no detectable activity of deoxythymidine kinase in the sponge connective tissue during the zinc restriction phase, but it was 0.385 units in Patient 1 and 0.321 units in Patient 2 after supplementation with zinc. The activities after supplementation became 70% of the mean normal values (14).

The effect of marginal zinc deficiency on gonadal functions were studied in 5 male volunteers (15). Their ages ranged from 51 to 65 years.

These subjects had normal gonadal function. They had no medical disorder, and they had taken no drugs prior to the studies known to affect testicular function.

Physical examination, routine laboratory tests, and chest roentgenogram were unremarkable. Testicular function as evaluated clinically and by the measurement of serum androgens, FSH, LH, and sperm count were normal in all patients. Zinc status as assessed by the determination of zinc concentration in plasma, erythrocytes, and hair was within normal limits.

During stabilization period, all subjects received hospital diet (10 mg of zinc intake daily). After stabilization, Subject 1 received the experimental diet providing 2.7 mg of daily zinc, with 10 mg of supplemental zinc (as zinc acetate) orally for 6 weeks. Subjects 2 and 3 received the experimentl diet providing 3.5 mg of daily zinc, with 30 mg of supplemental zinc (as zinc acetate) orally for 5 weeks. Subjects 4 and 5 were switched directly from the hospital diet to the experimental zinc-restricted diet. The experimental zinc-restricted diet provided 2.7 mg of daily zinc in Subject 1 for a period of 24 weeks; 3.5 mg of daily zinc in Subjects 2 and 3 for a period of 40 weeks; and 5 mg of daily zinc in Subjects 4 and 5 for a period of 40 and 32 weeks, respectively.

In Subjects 1, 2, and 3, the repletion phase was begun with the administration of 30 mg of zinc (as zinc acetate) orally while the same experimental diet was maintained for a period of

12 weeks in Subject 1 and 8 weeks in Subjects 2 and 3.   In
Subject 4, 10 mg of zinc (as zinc acetate) was mixed with the
experimental diet, for a period of 12 weeks. After the experimental diet, the hospital diet (10 mg of daily zinc) plus the
zinc supplementation were given for a period of at least 8 weeks
in these four subjects. Subject 5 was not given additional zinc
after his experimental diet was terminated but instead received
regular diet at home. The calories and protein intake of the
hospital and the experimental diet remained the same. Thereafter,
all patients received home diet similar to the hospital diet
providing 10 mg of daily zinc.

Blood samples were drawn for the determination of LH, FSH,
and testosterone before and after stimulation with GnRH. These
determinations were performed twice during stabilization period,
twice at the end of zinc-depletion period, twice during the first
6 months of zinc-repletion period, and twice after 6 to 12 months
of zinc repletion. Thus, the zinc repletion period was divided
into two phases: the early phase which reflected the first 6
months of repletion and the late phase which was extended beyond
6 months and up to 12 months of zinc repletion. Blood samples
were drawn from an indwelling intravenous catheter between 9:00
and 9:30 a.m. At least two baseline samples 30 min apart were
drawn for the determination of serum LH, FSH, and testosterone,
and then 200 µg of GnRH were injected intravenously. Blood
samples were drawn for the measurement of serum LH and FSH 15, 30,
60, 120, 180, and 240 min later. Blood samples were drawn for
the determination of serum testosterone 60, 120, 180 and 240 min
after GnRH injection. The baseline value of each hormone was
derived by calculating a mean of two to four baseline determinations. Means of two different determinations of GnRH-stimulation
tests were calculated for the stabilization, zinc-restriction,
and early and late phases of zinc-repletion periods. Thus
fluctuation of these hormones from hour to hour, as well as from
day to day, was avoided. Serum testosterone, LH, and FSH were
measured according to radioimmunoassay techniques.

Semen analysis was done every 4 to 12 weeks and after absti-
nence from sexual activity for 3 to 7 days before ejaculation.
Analysis was done within 30 min of ejaculation. Results of sperm
count were expressed in concentration of sperm per milliliter as
well as total number of sperm per ejaculate. A mean of two to
five sperm counts during each phase of the study were calculated.
Oligospermia was defined as a total sperm count less than 40
million per ejaculate.

Zinc concentrations in the erythrocyte and plasma decreased
significantly ($p < 0.01$) during zinc restriction in comparison
to the stabilization levels. During the early phase of zinc
repletion a slight increase in erythrocyte and plasma zinc levels
was noted, but these values were not statistically significant in
comparison to the zinc-restriction levels. A marked increase in
erythrocyte and plasma zinc concentration was observed during the

late phase of zinc repletion, and these levels were statistically significant (p < 0.005) when compared with the values of zinc restriction and early phase of zinc repletion.

Sexual hair and testicular size showed no change throughout the study period. However, sexual drive was diminished during zinc-restricted period as compared to the stabilization and zinc-repletion periods, but an objective assessment was unobtainable.

The baseline determinations of serum androgens and pituitary gonadotropins during stabilization period were normal. The mean ± S.E.M. of basal serum testosterone in all five subjects was 4.81 ± 0.78 ng/ml (normal 5.4 ± 0.3); serum LH, 14 ± 3.7 mIU/ml (normal 12.4 ± 2); serum FSH, 12.2 ± 2 mIU/ml (normal 10 ± 1). The range of sperm concentration in all five subjects was 131 to 825 millions/ml, with a mean ± S.E.M. of 305.9 ± 76.4. The range of total sperm count per ejaculate was 87 to 3925 millions, with a mean ± S.E.M. of 1114.7 ± 411.6 millions.

The sperm count declined slightly during zinc restriction and continued to decline in the early phase of zinc repletion. Oligospermia (total sperm count per ejaculate less than 40 million) was observed in four out of five subjects as a result of dietary zinc restriction. The mean ± S.E.M. of sperm concentration for the entire group was 59.2 ± 17.4 millions/ml, and the total sperm count per ejaculate was 53.6 ± 12.5 millions during the early phase of zinc repletion; the difference between these two values and those of the stabilization period were significant. During the late phase of zinc repletion, the sperm concentration increased to 207 ± 47.9 millions/ml, and the total sperm count per ejaculate increased to 416.3 ± 102.9 millions (mean ± S.E.M.). These values are considered to be within the normal range. There was a significant correlation between the sperm concentration and the total sperm count (4 = 0.90, p < 0.001). Changes in sperm motility and morphology throughout the study period were unremarkable.

The baseline serum testosterone decreased significantly during the early phase of zinc repletion and returned to normal levels during the late phase of zinc repletion. There was a slight decline in the maximal rise of serum testosterone after GnRH stimulation during zinc restriction and a more significant decline during the early phase of zinc repletion, with recovery to normal level during the late phase of zinc repletion. The changes of serum dihydrotestosterone were similar to those of serum testosterone, but statistically not significant.

Although the mean maximal rise in LH after GnRH stimulation was highest during early phase of zinc repletion, the values were not statistically significant as compared to the other periods. Baseline mean serum FSH was highest during zinc-restriction period, but again the values were not statistically significant as compared to the other periods.

Discussion

The purpose of our study was to produce only a mild zinc-deficient state in human volunteers, inasmuch as severe deficiency of zinc may be life-threatening, as seen in acrodermatitis enteropathica. Furthermore, it is the marginal deficiency of zinc that appears to be prevalent and likely to be missed clinically. Our data show that we were successful in producing a mild zinc-deficient state in human volunteers by dietary means. Zinc concentration of plasma, erythrocytes, leukocytes, and urine decreased when the daily intake of zinc was restricted to 2.7 and 3.5 mg in human volunteers. Changes in the activities of zinc-dependent enzymes such as alkaline phosphatase and ribonuclease in the plasma and thymidine kinase in the sponge connective tissue during the zinc restriction phase were also supportive of the conclusion that a zinc-deficient state was induced in the vonunteers. After supplementation with zinc, all the above mentioned variables returned toward normal levels. Inasmuch as all the other dietary constituents remained the same except for zinc intake, it is safe to presume that the above biochemical changes were due to zinc deficiency. Our zinc balance data show that we slowly depleted the body store of zinc by approximately 500 mg during the study period. Presuming the total body store of zinc in a 70-kg man to be around 1.5-2.5 g, 500 mg of zinc depletion represents approximately 30% or less of the body store. Thus the changes described here are those of a mild zinc-deficient state.

One unexpected finding was that the plasma ammonia level appeared to increase as a result of zinc deficiency. We have reported similar findings in zinc-deficient rats (16). This may have important health implications concerning zinc deficiency in man, because in liver disease hyperammonemia is believed to affect the central nervous system adversely.

In the first two subjects (Patients 1 and 2), the weight loss may have been partly due to a mild restriction in caloric intake in addition to a dietary zinc restriction. Also, that the synthetic diet based on soy protein may have been limiting in certain amino acids was considered. Because of these considerations, in the second group of subjects we increased the caloric intake, and in the base-line period of 5 weeks before zinc restriction, we supplemented these two subjects with 30 mg of zinc daily. The weight remained stable during the base-line period but decreased when the zinc was restricted subsequently. A slightly delayed and less marked response to zinc restriction in the second group of subjects was related to the fact that they received zinc supplement (30 mg daily) for 5 weeks before restriction, which may have built up the body store of zinc; furthermore, they received 3.5 mg of dietary zinc daily instead of 2.7 mg as was the case with the first two subjects. The amino acid content of the experimental diet was adequate according to the recommended dietary allowances, Food and Nutrition Board, National Research Council, National Academy of Sciences. Thus, the changes observed with respect to weight in our experiments were related to a dietary zinc restriction.

Our data suggest that the zinc-deficient state may have led to hypercatabolism of fat in our subjects. This is suggested by an increased fat loss and normal absorption of fat during the zinc restriction phase. In experimental animals an increase in free fatty acids has been observed as a result of zinc deficiency (17). Indeed, more studies are required in human subjects to document increased fat catabolism due to zinc restriction.

Direct measurement of DNA and protein synthesis in experimental animals suggest that in zinc deficiency protein synthesis is adversely affected (1, 2). Our own data on human subjects in these experiments show that the total protein, total collagen, and RNA-DNA ratio increased as a result of zinc supplementation. The activity of deoxythymidine kinase was not measurable during the zinc restriction phase but became 70% of normal level after supplementation with zinc for 3 months. Similar data have been published for experimental animals. Thus, our data show that deoxythymidine kinase in human subjects also is a zinc-dependent enzyme, and an adverse effect of zinc deficiency on this enzyme may be responsible for decreased protein synthesis. Our studies do not rule out an adverse effect of zinc deficiency on protein catabolism. Further studies are required to establish the effect of zinc restriction on protein catabolism.

Our data indicate that plasma alkaline phosphatase, plasma lactic dehydrogenase, and deoxythymidine kinase in sponge connective tissue in human subjects are zinc-dependent enzymes, inasmuch as changes in their activities were related to only one dietary manipulation, namely, zinc intake. Changes in the activities of plasma ribonuclease also appear to be related to zinc intake under the conditions of our experiments. Other trace elements may also inhibit its activity in vitro, but the effect of zinc is more pronounced. Previous investigations have related changes in the activities of plasma ribonuclease to protein intake (18); in these studies, changes in protein intake may have been associated with changes in plasma trace elements, thus accounting for alterations in ribonuclease activity. The measurement of the activities of these plasma enzymes may be helpful in correlating uncomplicated zinc status in humans, particularly if the changes are observed after zinc supplementation for a short period of time.

Changes in the plasma zinc concentration were observed within 4 to 6 weeks and correlated with the severity of dietary zinc restriction. Thus, plasma zinc may be very useful in assessment of zinc status in humans provided infections, myocardial infarction, intravascular hemolysis, and acute stress are ruled out (2). As a result of infections, myocardial infarction, and acute stress, zinc from the plasma compartment may

redistribute to other tissues, thus making an assessment of zinc
status in the body a difficult task. Intravascular hemolysis
would also spuriously increase the plasma zinc level inasmuch as
the concentration of zinc in the erythrocytes is much higher
than in the plasma.

Changes in the erythrocyte zinc concentration were slow to
appear as expected; on the other hand, changes in the leukocyte
zinc concentration appeared more sensitive to changes in zinc
intake. Urinary excretion of zinc decreased as a result of
dietary zinc restriction, suggesting that renal conservation of
zinc may be important for the homeostatic control mechanism in
man. Thus, measuring zinc concentration in a 24-h urine test
may be of additional help in diagnosing zinc deficiency provided
cirrhosis of the liver, sickle cell disease, and chronic renal
diseases are ruled out. These conditions are known to have
hyperzincuria and associated zinc deficiency.

Our data also indicate that during the zinc-deficient state
the subjects were in a more positive balance for zinc. This
would suggest that perhaps a test based on oral challenge of
zinc and subsequent plasma zinc measurement may be able to
distinguish between the zinc-sufficient and the zinc-deficient
state in human subjects. A study to test this possibility is
currently underway in our laboratory.

In our study, mild zinc deficiency as induced by dietary
zinc restriction has shown definite effect on gonadal function.
The sperm count was susceptible to dietary restriction of zinc.
Although there was only a slight decrease in the sperm count
during zinc restriction, oligospermia became significant during
the early phase of zinc repletion.

The nature of the delayed effect of zinc deficiency on
sperm count is not well understood. However, an explanatory
hypothesis can be drawn on the basis of other studies. The
developmental progression of spermatogenesis, from the origin of
spermatozoa in the germinal epithelium to mature spermatozoa, is
a prolonged process.   The duration of human spermatogenesis is
74 ± 4.5 days (19). Therefore, an insult affecting the germinal
cells may not be evident until several months later. The effect
of certain therapeutic drugs on spermatogenesis is an example.
In patients treated with cyclophosphamide, oligospermia or
azoospermia occurred several months after the treatment was
started (20-23). Similarly, spermatogenesis returned to normal
15 to 49 months after cyclophosphamide therapy was stopped. It
is not surprising therefore for an insidious insult to germinal
cells, such as zinc deficiency, to be manifested after such a
long period of time. If the temporal relationship between the
onset of injury to germinal epithelium and the sperm count is
true, one would expect an equal period of time to be required
for the recovery of spermatogenesis, once the ensuing disorder
has been corrected. In our subjects, the recovery of oligospermia occurred in similar fashion. The sperm count returned

to the baseline level after several months of zinc supplementation.

The long duration of human spermatogenesis, however, provides only a partial explanation for the effects of zinc on
gonadal function, inasmuch as we also observed a similar delayed
effect of dietary zinc restriction on plasma testosterone levels
in our subjects. Another contributory factor for the lack of
synchronism in testicular hypofunction with the periods of zinc
restriction and zinc repletion is perhaps the design of our
study. The slow induction of zinc deficiency by dietary means,
with its nutritional and metabolic consequences, required a relatively longer period of time to be reflected clinically on the
testicular function. Body stores of zinc were very slowly depleted and later were slowly repleted with zinc supplementation.
The dosage of supplemental zinc in these subjects was within a
physiologic rather than a therapeutic range. It is evident from
our data that repletion of body zinc store was not accomplished
until the late phase of zinc repletion which extended up to 12
months, and this correlated well with the observed effects on
testicular function in our study.

Acknowledgement

Supported in part by Sickle Cell Center Grant from National
Heart, Lung and Blood Institute, NIH, and USDA Competitive Research Grant.

Literature Cited

1. Prasad, A.S. "Trace Elements in Human Health and Disease";
       Academic Press, New York, New York, 1976, Chapter 1.
2. Prasad, A.S. "Trace Elements and Iron in Human Metabolism";
       Plenum Publishing, New York, New York, 1978, 251.
3. Abassi, A.A.; Prasad, A.S.; Ortega, J.; Congco, E.; Oberleas,
       D. Ann. Intern. Med. 1976, 85, 601.
4. Lei, K.Y.; Abbasi, A.; Prasad, A.S. Am. J. Physiol. 1976,
       230, 1730.
5. Hartoma, T.R.; Nahoul, K.; Netter, A. Lancet 1977, 2, 1125.
6. Prasad, A.S.; Oberleas, D.; Halsted, J.A. J. Lab. Clin. Med.
       1965, 66, 508.
7. Concon, J.M.; Soltess, D. Anal. Biochem. 1973, 53, 35.
8. Bauer, J.D.; Toro, G.; Ackerman, P.G. "Bray's Clinical
       Laboratory Methods," 6th ed., C.V. Mosby Co., St. Louis,
       Missouri, 1962, 333.
9. Rothstein, G.; Bishop, C.R.; Ashenbrucker, H.E. Blood, 1971,
       38, 302.
10. Bergmeyer, H.; Bernt, E.; Hess, B. "Methods of Enzymatic
       Analysis"; Bergmeyer, H., Ed.; Academic Press, New York,
       New York, 1963, 736.

11. Sekine, H.; Nakano, E,; Sakaguchi, K. Biochim. Biophy. Acta
     1969, 174, 202.
12. Prasad, A.S.; Oberleas, D. J. Lab. Clin. Med. 1973, 82, 461.
13. Fernandez-Madrid, F.; Prasad, A.S.; Oberleas, D. J. Lab.
     Clin. Med. 1973, 82, 951.
14. Prasad, A.S.; Rabbani, P.; Abbasi, A.; Bowersox, E.; Fox,
     M.R.S. Ann. Int. Med., 1978, 89, 483.
15. Abbasi, A.A.; Prasad, A.S.; Rabbani, P.; DuMouchelle, E.
     J. Lab. Clin. Med. 1980, 96(3), 544.
16. Rabbani, P.; Prasad, A.S. Am. J. Physiol. 1978, 235(2),
     E203.
17. Underwood, E.J. "Trace Elements in Human and Animal Nutri-
     tion"; Underwood, E.J., Ed., Academic Press, New York, New
     York, 1977, 196.
18. Albanese, A.A.; Orto, L.A. "Modern Nutrition in Health and
     Disease," 5th ed.; Goodhart, R.S.; Shils, M.E. Eds.; Lea
     and Febiger, Philadelphia, Penn., 1973, 28.
19. Heller, C.H.; Clermont, Y. Rec. Prog. Horm. Res. 1964, 20,
     545.
20. Kumar, R.; Biggard, J.D.; McEnvoy, J.; McGrown, M.G. Lancet
     1972, 1, 1212.
21. Qureshi, M.S.; Goldsmith, H.J.; Pennington, J.H.; Cox, P.E.
     Lancet 1972, 2, 1290.
22. Fairley, K.F.; Barnie, J.U.; Johnson, W. Lancet 1972, 1,
     568.
23. Buchanan, J.D.; Fairley, K.F.; Barnie, J.U. Lancet 1975,
     2, 156.

RECEIVED September 29, 1982

Trends in Levels of Zinc in the U.S. Food Supply, 1909–1981

SUSAN O. WELSH and RUTH M. MARSTON

U.S. Department of Agriculture, Human Nutrition Information Service, Consumer Nutrition Center, Hyattsville, MD 20782

The zinc level of the U.S. food supply has fluctuated between 11 and 13 mg per capita per day since
1909. These determinations, made for the first
time in 1980, were based on U.S. Department of Agriculture information: data on annual per capita food
use and data on the zinc content of food. Early in
the century, zinc was provided in almost equal pro-
portions by foods of animal and vegetable origin;
whereas, in recent decades, animal products have
provided approximately 70 percent. Three food
groups--meat, poultry, fish; dairy products; and
grain products--account for 75-80 percent of the
zinc; but, over the years, the proportion contributed by grain products has declined, and that
contributed by dairy products and especially by the
meat, poultry, fish group has increased.

The human diet has become a focal point of research in zinc
nutrition. In determining the zinc status of an individual, the
total dietary zinc level is not the only important factor. The
food sources of zinc and the presence of other substances in the
diet may profoundly affect zinc status. However, it is difficult
to estimate the proportion of an individual's total zinc intake
that is in a bioavailable form. The recommended dietary allowances (RDA) for zinc, which were set for the first time by the
National Academy of Sciences (NAS) in 1974 ([1]), are for the intake
of total zinc. These recommendations are based on the consumption
of a mixed American diet containing both animal and vegetable
products. Therefore, the total amount of zinc ingested by population groups should approximate the RDA, provided the diet does not
differ significantly from the diet considered by NAS to be typical.
Levels of zinc in the diets of American subjects have been
reported by several researchers. The studies shown in Table I
indicate the zinc content of food as consumed in the self-selected

Table I--Zinc levels in U.S. diets reported in
selected references

Dietary zinc level		Type of diet[a]	Refer-
mg/day	mg/1,000 kcal		ence
		Diets as consumed	
12.3	---	13 F; college	(2)
14.0	---	3 M; standard man study[b]	(3)
12.0	8.6	15 F; high school	(4)
12.6	8.8	33 F; college	(4)
10.1	5.6	18 M & 26 F; elderly	(5)
13.3	7.0	12 M & 7 F; college	(6)
8.6	4.2	11 M & 11 F; age 14-60	(7)
20.3	6.6	5 M; military, age 19-27	(8)
12.7	6.1	1 M & 40 F; college	(9)
		Other diets	
13.9	---	General hospital	(10)
11.4	4.4	General hospital	(11)
14.6	4.6	General hospital	(12)
11.0	4.1	50 college cafeterias	(13)
11.1	3.9	General hospital	(14)
19.7	5.4	Military cafeteria	(8)
18.7	4.7	FDA total diet study[c]	(15)

[a]"Normal" mixed diets of children over 10 years
of age and nonpregnant, nonlactating adults;
M = male and F = female.
[b]Composites prepared according to dietary records
of 2 M for 50 weeks and 1 M for 20 days.
[c]Composites represent diet of teenage boy (USDA
1965 Household Food Consumption Survey).

diets of free-living individuals and the zinc content of diets
planned for institutional feeding or research purposes. In these
studies, total zinc levels ranged from 8.6 to 20.3 mg per day.
The caloric levels of these diets varied considerably, partially
because of differences in experimental design. In general, zinc
intake increased as caloric intake increased, resulting in a more
narrow range for the zinc densities of the diets--4.1 to 8.8 mg
per 1,000 calories--than for the total zinc levels. In all of
the studies, the research methods involved chemical analysis of
dietary composites. Few estimates of zinc intake have been calculated from tables of zinc concentrations in food, even though
this approach is less complex and less costly. Calculation of
zinc intake from dietary survey data has been hindered by the
limited proportion of the many foods available in the market place
that have been analyzed for zinc.

The levels of zinc in the U.S. food supply, dating from 1909,
were estimated for the first time in 1980 (16). Calculations were
based on two types of information from the U.S. Department of
Agriculture (USDA): data on annual per capita food use and data on
the zinc content of food. This method is feasible because food
use in many instances is measured prior to final processing so
that the number of foods is relatively few and reliable zinc data
are available for these basic foods. Although food supply data do
not quantitatively measure food ingestion, they have the advantage
of being a consistent measure of nutrients provided by foods
available for consumption from the U.S. food supply. At the present time, this is the only method that has been used to evaluate
zinc levels in the American diet from a historical perspective.

Methods

Information on annual per capita food use in the U.S. is
obtained from published (17, 18) and unpublished data from the
Economic Research Service of USDA. Quantities are measured of
approximately 350 foods that "disappear" into the food distribution system. The amounts are derived by deducting data on ex-
ports, military use, year-end inventories, and nonfood use from
data on production, imports, and beginning-of-the year inventories. Because of the complexity of the food distribution system,
use of each food is not measured at the same point in the system.
Some foods are in a raw or primary state while others are retail
products when their use is measured. Subsequent losses that occur
in processing, marketing, and home use are not taken into consideration. Food supply data used in the preparation of this report
reflect, for the first time, revised population estimates for
1970-1981 based on the 1980 Census and revised estimates for fluid
milk and cream use from 1909-1981. Dairy product consumption data
for 1981 in this paper are preliminary.

The total zinc content of each food in the food supply is
based on values from provisional tables (19) and on more recent

unpublished information provided by the Nutrient Data Research
Group of the Consumer Nutrition Center of USDA. Although the
zinc content of some foods may have changed since the beginning
of the century, no data were available for early periods. Therefore, the most recent food composition data were used for the
entire time series. Average nutrient levels in the food supply
over spans of several years were used to represent nutrient intake
between 1909 and 1981, for example, 1909-13 and 1947-49. Estimates
of total nutrient levels in the food supply usually include quantities added in enrichment and fortification; however, such data are
not available for zinc. Additional information has been published
on methodology and other nutrients in the food supply (20).

Results

Since the beginning of the century, the zinc content of the
food supply has fluctuated between 10.6 and 12.8 mg per capita per
day (Table II). In general, zinc levels were at the lower end of
the range between 1920 and 1940. The level was lowest in the
1930's when economic conditions were depressed and consumption of
many foods, especially meats, was low. During the past decade,
zinc levels have been approximately the same as at the beginning
of the century, although since 1976, a slight downturn may be indicated. In addition, marked changes have occurred in the selection
of foods that provide zinc in the American diet.

Foods of animal and vegetable origin provided almost equal
amounts of zinc in the U.S. food supply until the mid-1930's
(Figure 1). As eating patterns changed, foods of animal origin
became more important sources of zinc. For the past two decades,
they have provided approximately 70 percent of the total zinc.
This shift in sources of zinc resulted from changes in contributions from three major food groups: (1) meat, poultry, fish (2)
dairy products, and (3) grain products. Together these groups have
accounted for about 75-80 percent of the total zinc in the food
supply (Figure 2). Since 1909-13, the meat, poultry, fish group
has remained the primary source of zinc. In recent years, it has
accounted for almost half of the total zinc; whereas, before 1947-
49, 38 percent or less of the total zinc came from this group.
The proportion of zinc contributed by dairy products also has increased over the years, making this group the second leading source
in the food supply. In recent decades, dairy products have contributed approximately 20 percent of the total zinc. On the other
hand, the proportion of zinc from grain products has decreased by
about one-half since the beginning of the century when it provided
27 percent of the total zinc. In the early 1940's, grain products
dropped from second to third place as a source of zinc in the
American diet.

Meat, poultry, fish group. Meat, which includes beef, pork,
veal, lamb and mutton, offal (edible organ meats) and game, has

Table II--Estimated zinc levels of the U.S.
food supply, 1909-13--81, per capita per day

Year	Zinc (mg)	Year	Zinc (mg)
1909-13	12.4	1973	12.2
1925-29	11.3	1974	12.3
1935-39	10.9	1975	12.3
1947-49	11.5	1976	12.8
1957-59	11.5	1977	12.6
1967-69	12.4	1978	12.4
1970	12.4	1979	12.3
1971	12.5	1980	12.2
1972	12.5	1981	12.2

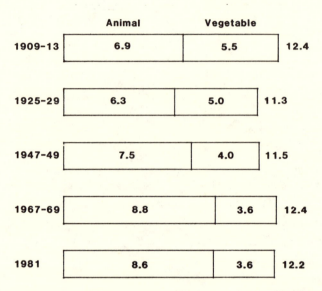

Figure 1. *Zinc from animal and vegetable sources in the U.S. food supply (mg/capita/day).*

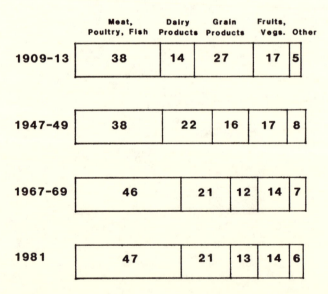

Figure 2. *Percentage of zinc from major food groups in the U.S. food supply. Potatoes, sweet potatoes, dry beans, dry peas, nuts, and soy products are included with fruits and vegetables. "Other" includes all remaining food groups. Components may not sum to total due to rounding.*

consistently provided about 80 percent of the zinc in this food
group (Table III). Therefore, meat is the primary source of zinc
in the meat, poultry, fish group as well as in the food supply as
a whole. Even when consumption was lowest, meat provided at least
27 percent of the zinc in the food supply. During the years of
high consumption from the mid-1960's through the 1970's, meat provided close to 40 percent of the zinc. The most recent data,
however, indicate a slight decline; between 1979 and 1981 meat
accounted for 37 percent of the zinc in the food supply.

Most of the changes in the amounts of zinc contributed by
meat can be related to the use of beef which accounts for the
largest single contribution of zinc in the American diet. Before
1953, consumption of beef was less than that of pork. However,
beef provided about twice as much zinc as pork because it is
higher in zinc concentration. Since the mid-1960's, beef has contributed three and sometimes four times as much zinc as does pork.
This change has been due to increased beef consumption which began
in the mid-1950's and peaked at 94 lb per capita in 1976. Between
1976 and 1980, beef consumption decreased 19 percent while pork
consumption increased 27 percent. However, 1981 data may indicate
another reverse in consumption trends. Pork contributed 1.0 mg of
zinc per capita per day or 8 percent of the zinc in the food supply
in 1981 while beef contributed 3.0 mg per capita per day or
24 percent of the total zinc.

Following beef and pork, offals make the next largest contribution of zinc among the meats. In general, they are very good
sources of zinc and, despite their low usage, they have consistently provided about 0.4 mg of zinc per capita per day. The
amounts of zinc contributed by offals and game have changed little
since the beginning of the century and the small amounts contributed by veal, lamb and mutton have declined slightly. Recently,
however, these meats have accounted for proportionately less of
the zinc from the meat, poultry, fish group due to the greatly
increased contribution from beef.

Poultry has become increasingly more important as a source of
zinc, providing about 7 percent of the total in the food supply
since 1979. In 1981, a total of 63 lb per capita of poultry was
consumed. Since the beginning of the century, use of chicken has
more than tripled and use of turkey has increased by a factor of
ten. Turkey is a more concentrated source of zinc than chicken;
therefore, even though use of turkey was only 21 percent as much
as use of chicken in 1981, turkey provided almost 27 percent of
the zinc from poultry. In the food supply as a whole, all poultry
contributed 0.9 mg of zinc per capita per day in 1981 compared
with 0.3 mg at the beginning of the century.

Fish, including finfish and shellfish, has contributed less
than half as much zinc in recent years as in 1909-13. This decrease occurred despite increases in consumption. A comparison of
contributions now and at the beginning of the century indicates
that more zinc is currently provided by fresh and frozen lean fish,

Table III--Zinc from the meat, poultry, fish group in the U.S. food supply, per capita per day

mg

Year	Meat					Poultry			Fish			Total[b]
	Beef	Pork	Offal	Other[a]	Total[b]	Chicken	Other[c]	Total[b]	Fin-fish	Shell-fish	Total[b]	
1909-13	1.9	0.9	0.4	0.4	3.7	0.2	d	0.3	0.2	0.6	0.8	4.7
1925-29	1.5	0.9	0.4	0.4	3.2	0.2	d	0.2	0.1	0.4	0.5	3.9
1947-49	1.9	0.9	0.4	0.4	3.6	0.3	0.1	0.3	0.1	0.3	0.4	4.3
1967-69	3.1	1.0	0.4	0.3	4.8	0.5	0.2	0.7	0.1	0.2	0.3	5.7
1975	3.4	0.8	0.4	0.2	4.8	0.5	0.2	0.7	0.2	0.2	0.3	5.8
1981	3.0	1.0	0.4	0.2	4.5	0.6	0.2	0.9	0.2	0.2	0.3	5.7

[a] Includes veal, game, lamb and mutton.
[b] Components may not add to total due to rounding.
[c] Includes turkey and small amounts of duck and geese.
[d] Less than 0.05 mg.

canned tuna, game fish, and shellfish other than oysters. Less
zinc is provided by canned salmon, sardines, cured fish, and
oysters. The overall decrease in zinc from fish can be attributed
almost entirely to the decline in the use of oysters, the most
concentrated food source of zinc. Oyster consumption has never
been large--a little more than 1 lb per capita per year in 1909-13;
but it was halved by 1935-39 and halved again by 1957-59. For the
past decade, oyster consumption has been about 0.2 lb per capita
per year. Consequently, the once significant contribution of zinc
from oysters declined from 0.6 mg per capita per day in 1909-13 to
0.1 mg in recent years.

Dairy products group. Zinc provided by the use of dairy
products has increased about 50 percent during the past 70 years
(Table IV). The largest amount of zinc contributed by dairy
products was 2.8 mg per capita per day in 1945 and 1946. These
were the years in which total dairy product consumption was the
highest. After a slight decline, the level of zinc from this food
group has remained relatively constant since the late 1940's.
Fluid milks and cream contributed heavily to the mid-century
peak for dairy product consumption. Use of fluid whole milk rose
until 1945, but thereafter declined markedly. At the beginning of
the century, whole milk was the most important source of zinc
among the dairy products, accounting for 63 percent of the total
from the group. However, by 1981, the proportion of zinc in the
dairy products group provided by fluid whole milk had declined to
26 percent. Lowfat milks have followed a different pattern. Use
of these products declined between 1909-13 and 1958; but since
then, use has almost quadrupled with most of the increase occurring in the last decade. Lowfat milks provided 0.5 mg per capita
per day or 20 percent of the zinc from the dairy products group
in 1981.
Processed milk products, which include condensed, evaporated,
malted and dry milks, cheese, whey, and ice cream and other frozen
desserts, account for a large part of the overall increase in zinc
from the dairy products group. Cheese with its markedly increased
use is chiefly responsible. Since 1909-13, the amount of zinc
provided by cheese increased more than fourfold and beginning in
1978, it has exceeded the amount contributed by fluid whole milk.
In 1981, cheese provided 0.9 mg of zinc per capita per day or
7 percent of all the zinc in the food supply. Use of ice cream
and other frozen desserts is now more than 10 times higher than
in 1909-13. Half of this increase occurred before the late 1930's,
and, since 1957-59, the small contribution of zinc from ice cream
and other frozen desserts has been relatively stable. Other processed milk products have accounted for more zinc in recent years
than at the beginning of the century. The amount, however, is
somewhat less than the 0.6 mg per capita per day provided in 1946
when consumption of these products was highest.

Table IV--Zinc from dairy products in the U.S. food supply, per capita per day

Year	Fluid milks and cream			Milk products				Total[b]
	Whole milk	Lowfat milks	Total[a],[b]	Cheese	Frozen desserts	Processed milks[c]	Total[b]	
				mg				
1909-13	1.1	0.3	1.4	0.2	d	0.1	0.3	1.7
1925-29	1.1	0.2	1.4	0.2	d	0.2	0.5	1.9
1947-49	1.4	0.2	1.6	0.4	0.1	0.5	0.9	2.6
1967-69	1.1	0.2	1.3	0.6	0.1	0.5	1.3	2.6
1975	0.8	0.4	1.3	0.7	0.2	0.4	1.2	2.5
1981	0.7	0.5	1.2	0.9	0.2	0.3	1.4	2.5

[a] Includes cream.
[b] Components may not add to total due to rounding.
[c] Includes evaporated, condensed, and dry milks and whey; excludes butter.
[d] Less than 0.05 mg.

Grain products group. Between 1909-13 and 1981, zinc from grain products decreased from 3.3 to 1.6 mg per capita per day (Table V). Use of these products declined by about 40 percent before the middle of the century. The decline continued, although less rapidly, until the early 1970's when an upward trend began due to increased use of wheat products and rice. Within the grain group, wheat products--white and whole wheat flour, semolina, and wheat cereal--have been the primary source of zinc accounting for about 60-70 percent of the total. Based on recent usage of wheat flour, if all were consumed in a whole grain form, wheat flour would provide about 5 mg of zinc per capita per day. For lack of better information, 1-2 percent of wheat flour in the food supply is considered to be consumed in a whole grain form. Therefore, zinc in the food supply from grain products may have changed more subtly than can be accounted for from available information. Nevertheless, food supply data show that the amount of zinc from wheat products has decreased by about one-half since the beginning of the century. At that time, they provided 2.0 mg of zinc per capita per day or 16 percent of the zinc in the food supply, whereas, in 1981, they provided 1.1 mg of zinc or 9 percent of the zinc in the food supply.

Corn products are the second leading source of zinc in this food group. In 1909-13, corn products provided 0.9 mg of zinc per capita per day, 30 percent of the zinc from grain products or 8 percent of all the zinc in the food supply. However, use of these products was halved by the mid-1930's and halved again by the early 1950's. Since the late 1950's, zinc provided by corn products has been relatively constant. Use of corn flour and meal alone has declined by 85 percent since 1909-13. For the last three decades, corn products have provided 0.2 mg of zinc per capita per day, or about 11 percent of the zinc in this food group.

Other grains such as rice, rye, barley, buckwheat, and oat products also contribute small amounts of zinc to the food supply. Among these grains, rice has comparatively high usage which has increased 64 percent between 1970 and 1981. However, because of the relatively low concentration of zinc in rice and the relatively low use of barley, buckwheat, oat and rye products, these grains have had little effect on the zinc level of the food supply.

Other food groups. Approximately 20-25 percent of the zinc in the food supply comes from four other food groups--(1) eggs, (2) dry beans, dry peas, nuts, and soy products, (3) potatoes and sweetpotatoes, and (4) vegetables (Table VI). In recent years, potatoes and sweetpotatoes have provided less than half as much zinc as at the beginning of the century due to greatly decreased use. Eggs and foods in the dry beans, dry peas, nuts, and soy products group are generally more concentrated sources of zinc than foods in the vegetable products group. Therefore, despite large differences in use among these three groups, they each provide roughly equal amounts of zinc which have remained relatively

Table V--Zinc from grain products in the U.S. food supply,
per capita per day

Year	Wheat products[a]	Corn products[b]	Rice	Other[c]	Total[d]
			mg		
1909-13	2.0	0.9	0.1	0.3	3.3
1925-29	1.7	0.6	0.1	0.4	2.7
1947-49	1.3	0.3	0.1	0.2	1.8
1967-69	1.0	0.2	0.1	0.2	1.5
1975	1.0	0.2	0.1	0.2	1.5
1981	1.1	0.2	0.2	0.2	1.6

[a] Includes white and whole wheat flour, semolina and cereal.
[b] Includes corn flour and meal, hominy, grits, cereal, and starch.
[c] Includes rye flour, barley products, buckwheat, and oat food products.
[d] Components may not add to total due to rounding.

Table VI—Minor sources of zinc in the U.S. food supply, per capita per day

Year	Eggs	Dry beans, peas, nuts, soya	Potatoes, sweet-potatoes	Vegetables	Fruits	Fats, oils[a]	Sugars, miscellaneous	Total[b]
				mg				
1909–13	0.4	0.5	0.9	0.5	0.2	0.1	0.1	2.7
1925–29	0.4	0.5	0.7	0.6	0.2	0.1	0.2	2.8
1947–49	0.5	0.5	0.6	0.6	0.2	0.1	0.2	2.8
1967–69	0.4	0.5	0.4	0.6	0.2	0.2	0.2	2.5
1975	0.4	0.6	0.4	0.6	0.2	0.2	0.2	2.5
1981	0.4	0.6	0.4	0.6	0.2	0.2	0.2	2.4

[a] Includes butter.
[b] Components may not add to total due to rounding.

constant over the years. Fruits, fats and oils, sugars, coffee and
cocoa contribute the remaining small proportion of zinc in the food
supply.

Discussion

     The primary purpose in estimating the nutrient content of the
U.S. food supply is to provide data which allow evaluation of
trends in food use and nutrient levels in the American diet since
the beginning of the century. However, the discrepancy that
exists between the levels of zinc in the food supply and NAS recommendations for dietary intake prompted further evaluation. To
determine whether or not the discrepancy is a cause for concern,
it is important to understand the inherent differences between the
methodology for setting RDA's and the methodology for estimating
the nutrient content of the food supply.
     Of major importance in a comparison of these data is the fact
that RDA's are for the amounts of nutrients actually ingested and
food supply estimates include some foods and nutrients that are
lost prior to ingestion. The amount of zinc that may be lost
during food processing, marketing, and home use is unknown, but it
is reasonable to assume that some loss occurs. If such losses
could be taken into consideration in food supply calculations,
undoubtedly the per capita levels of zinc would fall even further
below the RDA. Zinc may be consumed in dietary supplements or in
zinc-fortified foods, but the amount of zinc derived from such
sources is unknown.
     Nutrient density is considered an important factor in eval-
uating the adequacy of diets. However, this concept may be misleading when used with the food supply data because of the unknown
losses that occur after food use is measured. If the percentage
loss of a nutrient were different from the loss of food energy,
the nutrient density of the food supply would differ from the
nutrient density of the diet as ingested. In reference to zinc in
the food supply, there may be a selective retention of foods that
are high in zinc, such as lean meat, and discard of foods that are
high in calories, such as fat trimmed from meat and cooking oils.
Consequently, the zinc density of the food supply, which fluctuated
between 3.2 and 3.8 mg per 1,000 calories from 1909 to 1981, would
underestimate the zinc density of the U.S. diet as ingested,
although total zinc would be overestimated.
     RDA's are specified by sex/age categories, but they can be
expressed as a single value for the population by taking into consideration the distribution of sex/age groups within the United
States. On this basis, the "population" RDA for 1981 would be
14.2 mg per capita per day. This is still higher than the food
supply level of 12.2 mg per capita per day for 1981. Nutrient
levels in the food supply are averages for the entire population.
Therefore, certain segments of the population may choose from the
food supply a diet that provides the RDA, while others may choose

a diet that is low in zinc. In addition, when total dietary zinc is marginal or low, bioavailability may become a critical factor. Because of differences in the bioavailability of the zinc from foods of animal and vegetable origin, the RDA's are based on the consumption of a mixture of these foods. Although the food supply represents a mixed diet, individual diets may differ greatly from the average pattern.

"Recommended Dietary Allowances (RDA) are the levels of intake of essential nutrients considered, in the judgment of the Committee on Dietary Allowances of the Food and Nutrition Board on the basis of available scientific knowledge, to be adequate to meet the known nutritional needs of practically all healthy persons. RDA (except for energy) are estimated to exceed the requirements of most individuals and thereby to ensure that the needs of nearly all in the population are met. Intakes below the recommended allowance for a nutrient are not necessarily inadequate, but the risk of having an inadequate intake increases to the extent that intake is less than the level recommended as safe." (21) Few published dietary studies have reported zinc levels that meet the RDA either in terms of total zinc or zinc density. The results reported in this study indicate that the zinc content of the U.S. food supply has ranged between 10.6 and 12.8 mg per capita per day for the past 70 years. Differences in methodology do not diminish the discrepancy between RDA and food supply levels of zinc. Therefore, in relation to the RDA, present findings provide further evidence that the American diet is marginal in zinc.

Literature Cited

1. National Academy of Sciences, National Research Council, Food and Nutrition Board: "Recommended Dietary Allowances", 8th ed. Washington, DC, 1974.
2. Tribble, H.; Scoular, F.I. J. Nutr. 1954, 52, 209.
3. Tipton, I.H.; Stewart, P.L. Proc. 3rd Missouri Conference on Trace Substances in Environmental Health. 1970.
4. White, H.S. J. Am. Dietet. A. 1976, 68, 243.
5. Greger, J.L.; Sciscoe, B.S. J. Am. Dietet. A. 1977, 70, 37.
6. Haeflein, K.A.; Rasmussen, A.I. J. Am. Dietet. A. 1977, 70, 610.
7. Holden, J.M.; Wolf, W.R.; Mertz, W. J. Am. Dietet. A. 1979, 77, 23.
8. Milne, D.B.; Schnakenberg, D.D.; Johnson, H.L.; Kuhl, G.L. J. Am. Dietet. A. 1980, 76, 41.
9. Freeland-Graves, J.H.; Bodzy, P.W.; Eppright, M.A. J. Am. Dietet. A. 1980, 77, 655.
10. Gormican, A. J. Am. Dietet. A. 1970, 56, 397.
11. Osis, D.; Kramer, L.; Wiatrowski, E.; Spencer, H. Am. J. Clin. Nutr. 1972, 25, 582.
12. Brown, E.D.; McGuckin, M.A.; Wilson, M.; Smith, J.C., Jr. J. Am. Dietet. A. 1976, 69, 633.

13. Walker, M.A.; Page, L., J. Am. Dietet. A. 1977, 70, 260.
14. Klevay, L.M.; Reck, S.J.; Barcome, D.F. J. Am. Med. A. 1979,
     241, 18.
15. Harland, B.F.; Johnson, R.D.; Blenderman, E.M.; Prosky, L.;
     Vanderveen, J.E.; Reed, G.L.; Forbes, A.L.; Roberts, H.R.
     J. Am. Dietet. A. 1980, 77, 16.
16. Welsh, S.O.; Marston, R.M. Food Tech. 1982, 36, 70.
17. USDA, Economic Research Service. "U.S. Food Consumption,
     1909-63". Statistical Bul. No. 364, 1965.
18. USDA, Economic Research Service. "Food consumption, prices
     and expenditures". Statistical Bul. No. 672, 1981.
19. Murphy, E.W., Willis, B.W.; Watts, B.K. J. Am. Dietet. A.
     1975, 66, 345.
20. Welsh, S.O.; Marston, R.M. J. Am. Dietet. A. 1982, 81, 120.
21. National Academy of Sciences, National Research Council, Food
     and Nutrition Board. "Recommended Dietary Allowances", 9th
     ed., Washington, DC, 1980.

RECEIVED September 9, 1982

Assessment of the Bioavailability of Dietary Zinc in Humans Using the Stable Isotopes ^{70}Zn and ^{67}Zn

JUDITH R. TURNLUND

U.S. Department of Agriculture, Western Regional Research Center, Berkeley, CA 94710

JANET C. KING

University of California, Berkeley, CA 94720

Stable isotopes of zinc are valuable tools for studies of the bioavailability of dietary zinc in human subjects. Zinc occurs in nature as a mixture of five stable isotopes which are present in fixed ratios. ^{70}Zn and ^{67}Zn occur in the smallest amounts, 0.62% and 4.11%, respectively. Food sources of zinc which have been enriched with one of these isotopes are well suited for use as labels in zinc bioavailability experiments with analysis by thermal ionization mass spectrometry. Isotopes are preferred over balance methods for determining absorption of dietary minerals because specific feedings can be labeled, increasing the accuracy of results. Stable isotopes are advantageous in human subjects because no exposure to radioactivity is introduced. The method can be used safely even in infants and pregnant women. Bioavailability of dietary zinc was determined in six elderly men using stable isotopes and radioisotopes simultaneously. Stable isotope analysis was done using thermal ionization magnetic sector mass spectrometry, the method capable of the best precision and accuracy of stable isotope determinations. Zinc absorption from a semipurified formula diet was 17.6 ± 10.3% using stable isotopes. Mass spectrometer analysis agreed with radioisotopes absorption measurements, within the errors of the methods used. In another study, zinc absorption was compared in subjects consuming diets in which meat or plant foods were the primary source of protein. There was no difference in zinc absorption between the two groups.

Papers presented earlier in this symposium described a number of approaches which were used to determine bioavailability of dietary zinc. Experiments with animal models and human subjects were reported. Most scientists agree that as much information as possible should be first obtained from in vitro and laboratory animal experiments. However, since results of in vitro and animal experiments do not always agree with results of human experiments, research with human subjects is ultimately required to establish dietary requirements of humans and to determine bioavailablity of nutrients to humans.

Ethical and practical problems of research in humans severely restrict the types of experiments which can be carried out. Some of the approaches to determining bioavailability of zinc from foods to man include the following:

Approaches to determining zinc bioavailabilty in man.

Balance Studies. Examples of research using zinc balance were reported in this symposium. As these investigators and a recent review (1) pointed out, balance studies have limitations and must be augmented by other types of studies. Figure 1 (2) shows the components which contribute to balance determinations.

In our balance studies (3) we found that while mean zinc balance of a group of subjects appears to provide meaningful results, data of individual subjects is less reliable. While mean zinc balance in one study of elderly men consuming 15 mg of zinc per day was 0.1 mg/day, individual subjects appeared to retain or lose as much as 1.8 mg of zinc daily.

Numerous examples can be found in recent literature reporting marked positive and negative balance. For example, one recent paper contained a report of zinc balances which were as high as +11.6 mg/day (4). The effect of negative and positive balances at levels more commonly reported is shown in Table I. If we assume, for example, that balance of -3.9 mg/day continued indefinitely, an individual would lose his en-

Table I.   Examples of Reported Zinc Balance

Mg/Day	Mg/Year	% of TBC	Loss or Gain of TBC
-3.9	-1400	75	16 months
+5.6	+2100	115	11 months

Recommended Dietary Intake 15 Mg/Day
Total Body Content (TBC) 1400-2300 Mg

tire body zinc in 16 months. Zinc balance of +5.6 mg per day would result in significant accumulation and zinc stores would double in 11 months. While balance data may be useful for comparisons of groups or treatments when large differences are expected or when used in conjunction with other data, balance data alone must be interpreted with caution.

Zinc load tests. The appearance of zinc in the blood after a zinc load test has been used to detect differences in absorption between diets (5). Large differences in absorption can be detected with this method, but the absolute quantity of zinc absorbed or the percentage absorbed cannot be determined this way. In addition, the amount of zinc used in load tests is high compared to amounts usually consumed. These factors make it difficult to extrapolate from load test results to absorption of more usual zinc intakes.

Apparent absorption. When apparent absorption, or the amount of zinc fed minus the amount eliminated in the feces, is used to determine bioavailability, two problems must be considered. First, zinc collected includes non-absorbed dietary zinc, endogenous zinc from bile and pancreatic secretions into the small intestine, and sloughed intestinal cells, as shown in Figure 1 (2). In addition to these sources, intestinal mixing occurs and the dietary zinc consumed on one day is eliminated over many days. Figure 2 (6) shows the excretion pattern of one group of subjects, in which total excretion of unabsorbed ^{65}Zn took from 6 to 15 days. The fecal composites collected over a period of time required to completely eliminate the zinc fed on one day contained fractions of the unabsorbed dietary zinc from as many as 15 days prior to the feeding and fractions of dietary zinc consumed for up to 15 days following the feeding. As a result, complete collections of one day's dietary zinc contained not only that zinc, but fractions of zinc from 12 to 30 days' meals as well as fractions of endogenous zinc excreted over the same period of time. The zinc contributed by other meals and from endogenous sources can be ignored when using labels.

Radioisotopic labels. Radioisotopes of zinc, ^{65}Zn and ^{69m}Zn, have been used as labels in human absorption studies (7, 8) as described earlier in this symposium. Use of radioisotopes is limited because subjects are exposed to radioactivity. They should not be used in some population groups, such as infants and pregnant women. Further, the half life of ^{69m}Zn is too short for fecal isotope balance which requires collections over 12 to 15 days.

Stable isotopic labels. Stable isotopes of zinc can also be used as labels. The primary advantage of using stable iso-

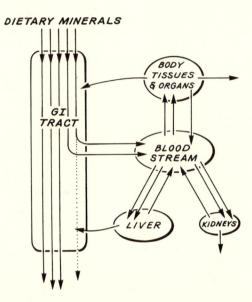

Figure 1.   Paths of dietary mineral absorption, distribution and mineral excretion. (Reproduced with permission from Ref. 2. Copyright Elsevier Publishing Co.)

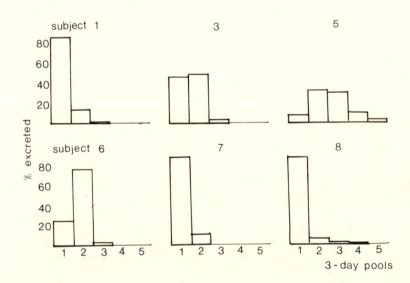

Figure 2.   Elimination patterns of labeled dietary minerals fed on 1 day and not absorbed. (Reproduced with permission from Ref. 6. Copyright 1982, American Society for Clinical Nutrition.)

topes is that there is no exposure to radioactivity when they
are fed. Stable isotopes can be used safely even in pregnant
women and in infants. Another advantage of stable isotopes
over radioisotopes is that they do not decay, so samples may
be stored indefinitely before analysis, and long-term studies
can be conducted. Most minerals have 2 or more isotopes which
occur in fixed ratios in all natural materials. Zinc has 5
isotopes, shown in Table II. Zinc enriched in any one of these
isotopes can be purchased from Oak Ridge National Laboratory.

Table II.   Stable Isotopes of Zinc

Isotopic Weight	Natural Atomic Abundance (%)
64	48.89
66	27.81
67	4.11
68	18.57
70	0.62

There are several disadvantages of using stable isotopes
as labels. They are more expensive than radioisotopes and the
supply is limited. Analysis requires extensive, time-consuming
sample preparation, and requires use of costly instruments not
readily available to all investigators. The cost and supply
of stable isotopes limits the quantities which can be used.
One mg of 99.7% ^{64}Zn costs \$0.45. The same purity of ^{70}Zn
cost \$155 per mg. The cost is expected to increase in the near
future, when a new batch of zinc isotopes becomes available.
Thus zinc isotopes for use in a single human study may cost
thousands of dollars. Specific isotopes are sometimes out of
stock and not available for several months or longer.

The analysis of stable isotopes requires extensive sample
preparation and sophisticated, expensive instruments. In our
work we use a thermal ionization mass spectrometer. Thermal
ionization mass spectrometers cost approximately \$300,000. In
contrast, analysis of the radioisotope ^{65}Zn requires little
sample preparation and analysis can be done quickly and easily
using a gamma counter.

Stable isotopes for research in man

History. Use of stable isotopes as research tools begin
in the 1930's with use of deuterium, ^{13}C and ^{15}N (9). In the
1940's radioisotopes became available. The advantages radioi-

sotopes offered, simple detection and little sample preparation, made them very attractive and they replaced stable isotopes as labels for most research purposes. Interest in stable
isotopes has been renewed and expanded rapidly in recent years.
Most work to date has been using ^{13}C, ^{15}N, and deuterium (9).
The first human experiment reported using a stable isotope of
zinc to determine zinc absorption was conducted in our metabolic unit (10). In that experiment, ^{70}Zn was fed to young
women to study the effects of oral contraceptives on Zn absorption. Neutron activation analysis was used for ^{70}Zn analysis.
No difference in zinc absorption was found due to oral contrac-
tive use. While it was apparent from this experiment that
stable isotopes would be valuable for zinc research in humans,
neutron activation analysis did not provide adequate precision
in this experiment to detect differences in absorption between
treatments. Differences between treatments would not have been
detected in a population of less than 100 subjects. Metabolic
units equipped for this type of experiment usually hold only 6
to 12 subjects. Also since these isotopes are expensive and
analysis is expensive and slow, studies using over 100 subjects
are not feasible. Therefore to obtain better analytical precision, we decided to attempt to use TI/MS for stable isotope
analysis.

Methods of stable isotope analysis. Three methods of a-
nalysis have been used for analysis of stable isotopes of minerals: neutron activation analysis (NAA) (10), gas-chromato-
graph-mass spectrometry (GC/MS) (11), and thermal ionization
mass spectrometry (TI/MS) (6). The maximum attainable precision of each of these analytical methods is compared in Table
III (2).

Table III. Maximum Attainable Precision
Of Stable Isotope Measurements

Method	Precision
	%
Neutron Activation Analysis	1-10
GC/mass spectrometry	2-100
Thermal ionization mass spectrometry	0.02-0.1

When GC/MS analysis is used, the metal of interest is
chelated to a volatile organic compound and analyzed. This
method is the fastest and least expensive method of analysis,
but it lacks the precision we require for reliable isotopic ratio determinations.

Because neutron activation analysis of zinc appeared to require little sample preparation and is capable of better precision than GC/MS, it was the method of analysis selected for our first experiment. Separation of zinc from other minerals in the samples was required, however, eliminating this as an advantage of NAA. Precision of this method was not as good in our experiment as had been expected.

Magnetic sector thermal ionization mass spectrometry is capable of far better precision and accuracy than the other two methods of analysis. A recent comparative study of NAA and TI/MS zinc analysis in human blood samples demonstrated the marked difference in precision between the two methods (12). The standard deviation of zinc ratio determinations was 0.13% for TI/MS and 8.3% for NAA. It is likely that, while maximum attainable precision for both methods is better than this (see Table III), these levels of precision are the best which can be achieved in biological samples. The excellent precision of TI/MS is offset by several disadvantages. Analysis is slow, requires considerable skill and experience, and is expensive. Extensive sample preparation is required to obtain pure samples pure enough for accurate analysis. However, the high degree of precision attainable by this method make it the only choice when excellent precision and accuracy (e.g. within 1%) are required.

Thermal ionization mass spectometery. In order to measure isotopic ratios of zinc by thermal ionization mass spectrometry, zinc must be separated from the rest of the sample in a very pure form. Organic materials are eliminated from samples by a combination of dry and wet ashing. Minerals are then dissolved in 8N HCl. Columns with a 8 mm diameter are filled with anion exchange resin which has been thoroughly washed with varying concentrations of ultra pure HCl to remove mineral contamination. After applying the sample to the column, it is rinsed with HCl to elute other minerals contained in the ash. Zinc elutes with 0.005 N HCl. Samples are purified further with two additional ion exchange column separations using 3 mm diameter columns. Zinc samples are then concentrated in teflon beakers for analysis by TI/MS.

Samples must be free of traces of organic compounds and of other ions which interfere with analysis. The oxides and fluorides of Ca, Cr, Fe, Ni, Ge, Sc, Ti and V, interfere with zinc determinations (6). Since naturally occurring zinc contains only 0.62% ^{70}Zn, 1 mg of natural zinc contains 6 µg of ^{70}Zn. Even 1 µg of an interfering element in the sample, 1/1000 of the total zinc, equal 15% of the ^{70}Zn and adversely affects ^{70}Zn analysis. Signal intensity from zinc is low compared to other interfering minerals like calcium. As a result calcium in amounts much less than 1 µg could interfere with zinc analysis.

The mass spectrometer we now use for zinc analysis, in the laboratory of Maynard Michel of Lawrence Berkeley Laboratory, is a thermal ionization mass spectrometer, a single direction focusing instrument with a 12" radius magnetic sector, double filament, rhenium ionizing source and electron multiplier detector. In addition, have done some preliminary work for Fe and Cu analysis with an automated TI/MS which speeds analysis considerably with excellent precision. We hope to be able to develop methods to use this automated instrument for zinc analysis as well.

Zinc absorption studies

Zinc absorption in elderly men. Zinc absorption was determined in a group of elderly men consuming a diet adequate in all nutrients (6). The diet was semi-purified with egg albumen as the protein source. Subjects lived in a metabolic unit for 83 days consuming these diets which contained 15 mg of zinc per day during the study. On day 28 of the study, 10 mg of ^{70}Zn was mixed into the formula diet in place of an equal amount of unenriched zinc. Zinc balance was determined in these subjects during a three week period of time which included the isotope feeding. On day 68, ^{70}Zn and radioactive ^{65}Zn were added to the formula diet and balance was determined again over a three week period of time (3). This experimental design permitted comparison of balance and absorption data in the same subjects on two different occasions under similar conditions, and a comparison of zinc absorption using both stable and radioactive isotopes. In addition, ^{70}Zn analysis was done using both NAA NAA and TI/MS.

Zinc absorption from the first feeding was 16.6 ± 3.3 (mean ± SEM) mg/day and from the seccond feeding was 18.0 ± 4.2 mg mg per day. Individual absorption measurements are shown in Table IV. There was no difference in absorption between the two periods. Absorption data determined by each of the three methods, radioisotopes, TI/MS and NAA, are shown in Table V. As was predicted, absorption determined by TI/MS analysis was in better agreement with absorption determined using radioisotopes than was absorption determined using NAA.

Zinc absorption from animal and vegetable protein diets. In a recent study we compared zinc absorption from different protein sources. Five young women were fed diets containing 70% animal protein for 21 days and 70% vegetable protein for another 21 days. The results of the study are shown in Table VI. There was no significant difference in zinc absorption between the two diets. Copper absorption was determined in this study also and did differ significantly between the two diets.

Table IV. Apparent Absorption of Zinc in
Elderly Men

	(% of Zn absorbed)	
Subject	MP 1	MP 2
1	9.0	−0.8
2	13.5	
3	22.8	29.1
5	25.8	25.1
6	11.7	18.9
7		16.5
8		18.9
Mean ± SEM	16.6 ± 3.3	18.0 ± 4.2

Table V. Apparent Absorption of Zinc in Six
Elderly Men

	Radioisotopes	Stable Isotopes	
		TI/MS	NAA
	%	%	%
Zn	12.4 ± 8.7	18.0 ± 10.3	27.7 ± 11.4*

* Mean of 4 subjects
Mean ± SD

Table VI. Zinc Absorption in Young Women

Subject	Animal Protein Diet	Plant Protein Diet
	(% Absorbed)	
A	21.5	24.8
B	25.3	29.6
C	30.9	21.1
D	19.4	22.4
E	21.9	29.1
Mean ± SD	23.8 ± 4.5	25.4 ± 3.8

We have used stable isotopes in several other zinc absorption studies. Sample analysis is in progress from a study in which ^{67}Zn was used to determine the effects of phtate and cellulose on zinc absorption in man. Studies were also conducted to determine zinc absorption from vegetarian diets, the effect of vitamin B-6 status on zinc absorption, and the effect of level of dietary zinc on zinc and copper absorption.

Other key issues in zinc absorption which will be studied with stable isotopes are the bioavailability of zinc to human infants from breast milk and formulas and comparison of the availability of intrinsic and extrinsic zinc labels.

Literature Cited

1. Kelsay, J.L., Cereal Chem. 1981, 58, 2-5.
2. Turnlund, J.R., Sci. of Total Environ., 1982. In press.
3. Turnlund, J., Costa, F., and Margen, S. Amer J Clin Nutr. 1981, 34, 2641-2647.
4. Obizobo, I.K., Nutr Rep Int. 1981, 24, 203-210.
5. Solomons, N.W., Jacob, R.A., Pineda, O., and Viteri, F.E., J. Nutr. 1979, 109, 1519-1528.
6. Turnlund, J.R., Michel, M.C., Keyes, W.R., King, J.C., and Margen, S. Amer J Clin Nutr. 1982, 35. 1033-1040.
7. Aamodt, R.L., Rumble, W.F., Johnston, G.S., Markley, E.J. and Henkin, R.I., Amer J Clin Nutr 1981, 34: 2648-2652.
8. Arvedson, B., and Cederblod, A. Int. J. Nucl Med and Bio. 1978, 5:104-109.
9. Klein, P.D., Hachey, D.L., Kreek, M.J., and Schoeller, D.A. In. T. A. Baillie. (ed). "Stable Isotopes: Applications in Pharmacology, Toxicology and Clincal Research" University Park Press, Baltimore, 1978, p. 3-14.
10. King, J.C., Raynolds, W.L. and Margen, S., Amer J Clin Nutr. 1978, 31, 1198-1203.
11. Hachey, D.L., Blais, J.C., and Klein, P.D., Anal Chem. 1980, 52:1131-1135.
12. Janghorbani, M., Young, V.R., Gramlich, J.W. and Machlan, L.A., Clin Chem Acta. 1981, 114:163-171.

RECEIVED September 3, 1982

Stable Isotope Approaches for Measurement of Dietary Zinc Availability in Humans

MORTEZA JANGHORBANI, NAWFAL W. ISTFAN, and
VERNON R. YOUNG

Massachusetts Institute of Technology, Nuclear Reactor Laboratory,
Department of Nutrition and Food Science, and Clinical Research Center,
Cambridge, MA 02139

Direct measurement of dietary zinc availability in
humans requires development of the stable isotope
tracer methodology. Several aspects of this inte-
grated methodology are considered and briefly discussed. These are: analytical isotopic measurement
methodology, consequences of the finite precision of
isotopic measurements, validation of in vivo measurements, and several aspects of biological labeling of
human foods. It is shown that Radiochemical Neutron
Activation Analysis provides a suitable method for
accurate measurement of the stable isotopes ^{64}Zn,
^{68}Zn, and ^{70}Zn. Employing simultaneous administration of the stable isotope ^{70}Zn and the radiotracer
^{65}Zn in rats, it is shown that the two methods yield
identical results. The issue of exchangeability of
an "extrinsic tag" with food zinc is addressed and
data, using ^{68}Zn-labeled chicken, are given to show
the high correlation between the results from the
two labels. Some applications exploring the nature
of diet protein with respect to zinc absorption are
reviewed.

There have been few direct measurements in man of the availability of dietary zinc and, in consequence, there is limited
knowledge of the quantitative effects of nutritional factors on
the absorption of zinc from foods and diets as consumed by human
subjects. Much of the published data have been obtained in
studies involving the chemical balance method (1). Better
methods and more complete information about the availability of
dietary zinc has become important, in part, due to the suggestion
that marginal zinc insufficiency may be more widespread in various populations than earlier thought to be the case (2). In
addition, the introduction of new foods and the possible effects
of food processing must be assessed for their effect on zinc
nutriture, further emphasizing the need to generate more complete

information about the availability of zinc from the foods consumed by human populations.

The chemical balance technique is inadequate for the accurate assessment of dietary zinc availability and it does not permit, in any case, an evaluation of zinc availability from the
separate components of a composite meal. Furthermore, although
measurement of changes in the concentration of plasma zinc (3)
following ingestion of pharmacological doses of the metal, has
been used in studies of dietary zinc availability, this approach
has not been validated with regard to its reliability and it
cannot provide absolute data on the absorption of zinc. Indeed
the only viable approach that can be followed is one involving
use of a suitable isotopic zinc tracer.

The available isotopes of zinc are given in Figure 1 (both
stable and radioactive). Thus, two radioisotopes (^{65}Zn, t 1/2 =
245d; ^{69m}Zn, t 1/2 = 13.8h) of the metal are available as tracers.
Of these, ^{69m}Zn possesses a short half-life (13.8h) and, except
for relatively short-term measurements involving whole-body
counting or plasma appearance studies (4), its application in
human metabolic studies is limited. For isotope balance studies,
requiring a minimum period of about a week, ^{69m}Zn is obviously
unsuitable. The other radioisotope of zinc (^{65}Zn) possesses a
long physical half-life (245 days) and, theoretically, is suitable
for use in human metabolic studies. However, its long biological
half-life, $\geqslant$500 days (3), introduces problems of human exposure to
ionizing radiation. Furthermore, use of radiotracers is, of
course, not warranted in children and women of childbearing
status, groups for whom the least is known about zinc metabolism
and dietary zinc availability. In addition to these concerns,
use of radiozinc in studies requiring biological labeling of
foods, that may then need processing and preparation, adds further complexity in the conduct of human studies.

As seen from Figure 1, natural zinc consists of five stable
isotopes. The most abundant is ^{64}Zn, with ^{70}Zn in the lowest
abundance. These isotopes can be obtained as highly enriched
formulations and this also is shown in Figure 1. Furthermore,
stable isotopes of zinc can be used in a manner analogous to
their radioisotopes for tracer studies, but important methodological and analytical differences arise with the use of these
forms of isotopic zinc. It is our purpose to discuss the present
state of development of the stable isotope tracer approach, specifically in relation to its application in human studies concerned with zinc availability from foods and diets. We will
present data, obtained in our laboratories, that emphasize the
feasibility and potential of this novel technique. To begin with,
however, a brief discussion of some of the concepts and analytical considerations that apply generally to stable isotope
methods should be made.

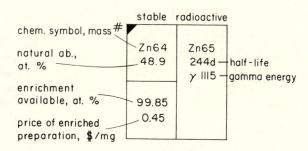

Zn64	Zn65	Zn66	Zn67	Zn68	Zn69m	Zn70	Zn7lm
48.9	244d	27.8	4.1	18.6	139h	0.62	4.lh
	γ1115				γ439		γ386
99.85		98.22	93.11	99.0		65.51 99.72	
0.45		0.65	4.15	1.35		23.10 154.9	

stable   radioactive

chem. symbol, mass #

Zn64   Zn65

natural ab.,   48.9   244d — half-life
at. %   γ 1115 — gamma energy

enrichment
available, at. %   99.85

price of enriched   0.45
preparation, $/mg

Figure 1.   Isotopic constitution of zinc. (Reproduced with permission from Ref. 9.)

Some Concepts and Analytical Considerations

For the present discussion, we define dietary zinc availability as the difference between intake and fecal excretion of zinc of dietary origin. This definition is, thus, the same as that for true absorption as given by Weigand and Kirschgessner (5). It explicitly excludes the fecal zinc of endogenous origin, but does not necessarily take into account re-entry of newly absorbed dietary zinc into the intestinal tract that is subsequently eliminated in the fecal zinc pool. Furthermore, depending upon the experimental approach, an overestimation of the fraction of dietary zinc that is available to the general circulation may arise due to an initial retention of zinc in the intestinal tissues and/or the liver (3). However, the present definition permits an approach that lends itself to accurate experimental quantitation of zinc absorption, using the method of stable isotope probes. Since it is common practice to express absorption of zinc as a fraction of the intake, the following expression applies:

$$F = \frac{I - 0}{I} \qquad (1)$$

where: F = fractional absorption of food zinc
        I = intake of food zinc during a defined time period
        0 = fecal output of <u>food</u> <u>zinc</u> corresponding to the
              time period for I

Measurement of fecal zinc does not yield the value 0, but rather the total zinc excreted (of dietary and endogenous origin). Thus, determination of total zinc in the diet and the fecal pool will not provide an estimate of the absorption of zinc. However, if the isotopic makeup of the dietary zinc is suitably altered, via administration of a labeled diet, then isotopic analysis of the diet and of the corresponding fecal pool will satisfy the requirements of the above definition. For the simplest case where the natural isotopic composition of a single component of the diet has been altered (single-labeling) an estimate of the fractional absorption (F_1) of the mineral from the labeled diet component, can be made from two simultaneous equations, as follows (See Table I for definitions):

$$A_{f,1} = A^*_{0,1,1} (1 - F_1) + A_{e,1} \qquad (2)$$

$$A_{f,2} = A_{0,2,1} (1 - F_1) + A_{e,2} \qquad (3)$$

$$\text{and } R_{1/2} = A_{e,1}/A_{e,2}$$

TABLE I

Definitions of Terms Used in Equations (2)-(7)

F_j	Fractional absorption of zinc from the labeled diet pool j.
$A_{o,j,k}$	Amount of isotope j present in diet pool k. If $A_{o,j,k}$ is super-scripted ($A^*_{o,j,k}$), it indicates that isotope j has been enriched in diet pool k.
$A_{e,j}$	Amount of isotope j in the fecal pool originating from unenriched sources.
$R_{j/j'}$	Natural mass isotope ratio for isotope j to j'.
$A_{f,j}$	Amount of isotope j in fecal pool of interest.
σ_x	Standard deviation associated with value of x.
n	Number of subjects in an experimental population.

Solution of these two equations yields the appropriate expression
for F_1; fractional absorption of the mineral from the labeled
component of the diet:

$$F_1 = \frac{A^*_{0,1,1} - A_{f,1} - R_{1/2} \cdot A_{0,2,1} + R_{1/2} \cdot A_{f,2}}{A^*_{0,1,1} - R_{1/2} \cdot A_{0,2,1}} \tag{4}$$

In the simple case for $A_{0,2,1} \cong 0$, i.e. labeled diet component
almost exclusively contains the isotope 1, Equation (4) can be
simplified to:

$$F_1 = \frac{A^*_{0,1,1} - A_{f,1} + R_{1/2} \cdot A_{f,2}}{A^*_{0,1,1}} \tag{5}$$

Equation (4) above is the generally applicable expression
for a singly labeled diet component when this component also
contains a measurable level of isotope 2. For the case where the
labeled component has been arranged so that its content of isotope 2 is negligible, as in the case of giving an oral dose of a
highly enriched solution, as in the extrinsic tag approach, then
Equation (5) is applicable. For situations where more than one
component of the diet has been labeled by appropriate use of a
selection of different stable isotopes, a more complex set of
multiple simultaneous equations can be described (6) for determination of the absorption of zinc from the various labeled diet
components.
      Thus, the first point to be made is that in the calculation
of F_1 from Equation (4) values of four parameters must be known
accurately ($A^*_{0,1,1}$, $A_{0,2,1}$, $A_{f,1}$ and $A_{f,2}$). Since these four
parameters are measured quantities, each with a finite value of
uncertainty, the calculated value of F_1 will also contain a
resultant value of uncertainty. The magnitude of this "measurement uncertainty" is of fundamental concern in the accurate
determination of F_1 and, initially, arises solely from the uncertainties of measurement in the four parameters. In addition,
other sources of variability, such as those associated with the
incomplete recovery of unabsorbed isotopes via fecal excreta or
inter- and intra- individual differences in zinc absorption,
contribute to the precision of the determination of zinc availability. It is important to evaluate the quantitative significance of this "measurement uncertainty" in order to evaluate the
requirements and practical feasibility of alternative stable
isotope approaches. The magnitude of this "measurement uncertainty" is obtained from the expression:

$$\sigma_{F_1}/F_1 =$$

$$\sqrt{\frac{\sigma^2_{A^*_{0,1,1}} + \sigma^2_{A_{f,1}} + R^2_{1/2} \cdot \sigma^2_{A_{0,2,1}} + R^2_{1/2} \cdot \sigma^2_{A_{f,2}}}{(A^*_{0,1,1} - A_{f,1} - R_{1/2} \cdot A_{0,2,1} + R_{1/2} \cdot A_{f,2})^2} + \frac{\sigma^2_{A^*_{0,1,1}} + R^2_{1/2} \cdot \sigma^2_{A_{0,2,1}}}{(A^*_{0,1,1} - R_{1/2} \cdot A_{0,2,1})^2}} \quad (6)$$

If measurements are made with several (n) subjects, then

$$\sigma_{\bar{F}}/\bar{F}_1 = (\sigma_{F_1}/F_1)/\sqrt{n} \tag{7}$$

A second point to be emphasized concerns the extent to which there is exchange of isotope among the dietary pools (6). Experimental investigation of these issues requires biological labeling of food models with stable isotopes and subsequently the simultaneous administration to human subjects of diets with more than a single labeled pool (7). In such experiments the simplifying assumptions of highly enriched pools (as in the case of an "extrinsic tag") are usually not valid so that set up of exact expressions are important. For the case of a diet containing two enriched components, pools 1 and 2, each enriched with respect to a distinct isotope but necessarily containing all the other stable isotopes, the following set of three simultaneous equations is applicable:

$$A_{f,1} = A^*_{0,1,1} (1 - F_1) + A_{0,1,2} (1 - F_2) + R_{1/3} \cdot A_{e,3} \tag{8}$$

$$A_{f,2} = A_{0,2,1} (1 - F_1) + A^*_{0,2,2} (1 - F_2) + R_{2/3} \cdot A_{e,3} \tag{9}$$

$$A_{f,3} = A_{0,3,1} (1 - F_1) + A_{0,3,2} (1 - F_2) + A_{e,3} \tag{10}$$

Again the measured quantities are the nine isotopic contents of the two diet pools and the fecal pools (of the three isotopes 1, 2, and 3) and the unknowns are the three quantities F_1, F_2, and $A_{e,3}$. The mathematical expressions given above are important since, as we will see further below, they will permit determination of the feasibility of various stable isotopes in human studies and help establish the optimum design of metabolic studies based on stable isotopes. As we will see later, the conclusions derived from these expressions are both important to the proper design of metabolic studies and are in fact validated by experimental data derived from actual metabolic studies.

Methodological Issues

Some of the important methodological issues that should be considered when stable isotopes and the isotope balance method are to be used for measurement of dietary mineral availability are:
1. Methods of isotopic measurement
2. Consequences of finite precision of measurement
3. Validity of in vivo measurements
4. Methodology of biological labeling
We have begun to examine these various issues and although our studies are still at an early' stage we will review briefly our findings and experience to date.

Methods of Isotopic Measurement. Two methods of stable isotope measurement are applicable: 1) Mass Spectrometry, and 2) Neutron Activation Analysis. The former is described elsewhere in this volume and so we will provide a brief description of the latter method. We have given more detailed discussions elsewhere (8, 9).

Of the five stable isotopes of zinc, only three (^{64}Zn, ^{68}Zn, and ^{70}Zn) can be measured with (delayed gamma) Neutron Activation Analysis and high-resolution gamma spectrometry. The required methodology for the measurement of ^{70}Zn is the most stringent of the three isotopes and the necessary Radiochemical Neutron Activation Analysis has been developed and described in detail elsewhere (8). The salient features of neutron activation of these three isotopes are tabulated in Table II and from these, two points emerge: First, with the availability of thermal neutron flux of about 5×10^{13} n cm^{-2} s^{-1} and sample irradiation time of about 10 minutes, the three isotopes can be measured simultaneously on the same sample. Second, precision of the measurement based on neutron activation cannot be improved beyond that determined by the "counting statistics" (Similar to signal/noise ratio for other measurements). Thus, if the "counting statistics" of a certain gamma line limits the measurement precision to 1% relative value, the overall isotopic measurement can only approach but not exceed this value, depending on the magnitude of uncertainties of other steps in the overall analytical procedure. It is generally accepted that absolute isotopic measurements can be carried out by neutron activation analysis with relative precision in the range of 1-10% (10). Thus, we (8) have found that overall measurements of ^{64}Zn, ^{68}Zn, and ^{70}Zn can be carried out in unenriched human stools with analytical precisions of about 5% for ^{70}Zn and with values approaching 1% for the other two isotopes. In stools suitably enriched with ^{70}Zn, overall analytical measurement precision of 1% can be approached for all three isotopes (8).

Consequences of Finite Precision of Measurement. The available measurement methodology indicates that overall isotopic measurement in human fecal samples suitably enriched with ^{70}Zn can be carried out routinely with relative precisions of 1-5%, depending on the specific experimental procedure and the effort taken to obtain a high level of precision. It is important to examine the consequences of this finite measurement precision in relation to the dose of oral tracer or degree of dietary enrichment and expected measurement precision value associated with the calculated value of F. Again this has been discussed in detail previously (8), and we summarize here their salient features by reference to the data given in Tables IIIA and IIIB.

These data have been calculated using Equation (6) above and the dietary conditions assumed are those noted in the table. They indicate several basic points. First, the value of the parameter σ_F/F is dependent on the actual magnitude of F, the overall measurement precision of each of the required isotopic measurements, and the level of dietary enrichment. Optimization of the experimental design with respect to these parameters is important not only from the point of view of accuracy but also isotope cost. In general, at any level of precision and dietary enrichment, the value of σ_F/F decreases exponentially as F is increased, approaching an asymptote (8). As measurement precision is improved the value of this parameter is also reduced. Similarly, increasing dietary enrichment improves on this parameter. For example, at 30% absorption, improving the overall measurement precision from 5% to 1% (Table IIIA) would reduce the value of σ_F/F from .27 to .053 when a 1.0 mg dose of ^{70}Zn is used. On the other hand, increasing the administered dose from 1.0 mg to 10.0 mg ^{70}Zn would only reduce the value of this parameter marginally, but with a large increase in isotope cost. Currently a 5% measurement precision is readily achieved. Thus, the expected measurement component of variability for a mean zinc availability of 30% in a group of nine subjects would amount to ± 0.09. Of course, the experimentally observed value would be larger than this, depending on the other components of variability, as mentioned earlier.

Administration of about 1 mg of ^{70}Zn would appear to be acceptable compromise in studies with adult subjects consuming about 15 mg/day of natural zinc. However, it might be necessary, or desirable, to enrich the diet or a test meal with a second isotope (6, 7). Since ^{67}Zn cannot be measured by neutron activation, ^{68}Zn is a potential candidate for this second tracer. However, the natural abundance of this isotope is high (18.6 atomic %) so that the feasibility of its application at physiological intake levels must first be established. In situations where ^{68}Zn is used as a label, ^{64}Zn serves as the reference isotope because the precision of measurement of ^{64}Zn is better than for ^{70}Zn.

We have summarized in Table IIIB the relevant data where ^{68}Zn may be used as the label. Again, assuming total zinc intake is 15 mg daily of natural zinc, we arrive at the values for σ_F/F given in columns 2 and 3 of the table. For example, if F = 0.30 and measurement precision is 5% then the value of σ_F/F would be

0.70 (cf. 0.27 for ^{70}Zn as label). If, on the other hand, the
basal zinc of the diet could be given largely in the form of ^{66}Zn,
in order to reduce the contribution made by dietary sources to
^{68}Zn in the fecal pool, then the fecal ^{68}Zn would originate from
endogenous sources. Assuming the latter to be approximately 4 mg
zinc daily, then at F = 0.30 and measurement precision of 5%, the
value of σ_F/F would be 0.30 or similar to that predicted for ^{70}Zn
as label. In addition, if the precision of measurements can be
improved then the values of this parameter would be further reduced. Based on these considerations, it would appear that both
^{68}Zn and ^{70}Zn are potentially suitable isotope probes for assessment of dietary zinc availability in adult subjects. Furthermore,
use of both ^{68}Zn and ^{70}Zn in double labeling studies, using ^{64}Zn
as the reference isotope, is quite feasible.

TABLE II
Selected Parameters of Neutron Activation of Stable
Isotopes ^{64}Zn, ^{68}Zn, ^{70}Zn of Interest to Isotopic
Fecal Balance

Nuclear Rx	Half-life of Product	Gamma Energy for Most Intense Line (kev)	Expected Precision Based on Counting Statistics (%)*
^{64}Zn(n,γ)^{65}Zn	245 days	1115	∿1
^{68}Zn(n,γ)^{69m}Zn	13.8 hours	439	<.3
^{70}Zn(n,γ)^{71m}Zn	4.1 hours	386	< 5

*Values are for fecal sample size containing about .2 mg
unenriched zinc. For other experimental conditions (neutron flux, etc.), see ref. (8).

TABLE IIIA
Effect of Dose Level and Isotopic Measurement Precision on
the Expected Value of σ_F/F Using ^{70}Zn as Label*

Fractional Absorption	Expected Value of σ_F/F for Given Values of Dose and Precision			
	1 mg ^{70}Zn Dose		10 mg ^{70}Zn Dose	
	1%	5%	1%	5%
.10	.17	.86	.14	.69
.30	.053	.27	.043	.21
.50	.030	.15	.025	.12

Conditions: 5 day fecal pool contains 65. mg unenriched Zn
(15 mg/day Zn intake, 13 mg/day fecal zinc loss);
^{70}Zn is given at 100% enrichment, ^{68}Zn is the
second measured isotope.

*For subject population of n, the expected values given in the
table are multiplied by $1/\sqrt{n}$

TABLE IIIB
Effect of Selected Experimental Parameters
on σ_F/F Using ^{68}Zn as Label[*]

Fractional Absorption	Expected Value of σ_F/F for Given Conditions of Zinc Supplment and Precision			
	Natural Zn intake[a]		^{66}Zn Intake[b]	
	1%	5%	1%	5%
.30	.14	.70	.059	.30
.50	.081	.41	.035	.18

Conditions: 5 day fecal pool contains 65.0 mg of total unenriched Zn of which (a) 12.5 mg is ^{68}Zn or (b) 4.04 mg is ^{68}Zn from endogenous sources; dose of ^{68}Zn label is 5.0 mg at 100% enrichment.

[*]For subject population of n, the expected values given in the table are multiplied by $1/\sqrt{n}$

Validity of in vivo Measurements. Since the stable isotope balance approach is a novel method, its validity needs to be established, and the problems that must be included in this examination are briefly discussed below.

a. Excretion of Label via Feces. Two issues are important here. First, since the method is a balance technique, it must be insured that all of the unabsorbed dietary label is included in the fecal pool. This requires establishment of the transit time for the unabsorbed label. Some representative data for subjects consuming a soy-protein based, liquid-formula diet and who received six sequential doses of ^{70}Zn solution in their three daily meals over a two-day period are given in Figure 2 and more extensive data given elsewhere (11). The data illustrate that under these conditions all of the unabsorbed isotope was excreted within no more than five consecutive stools passed after the first administration of the isotope. Second, following absorption of the label, the endogenous secretions of zinc will be labeled. Thus, it should be determined whether enrichment of endogenous zinc in feces can be experimentally observed and, if so, what the effect would be on the estimate of F. The data presented in Figure 3 indicate that the reentry of isotope cannot be detected within the precision of our available neutron activation methodology. Taking the data of Spencer et al. (12) and assuming that 0.20 - 0.45% of the absorbed dose is reexcreted daily in the stools, it can be calculated, depending on the actual value of F, that reentry of absorbed label should increase the natural abundance of ^{70}Zn in stools by approximately 1-2% (for F = 0.50, enrichment should be 1.0 - 2.3% mass isotopic excess over the next 40 days). If the measurement methodology possessed a precision of much better than 1% (e.g. 0.1% as might be achievable with Isotope Ratio Mass Spectrometry) this reentry could be experimentally measured. Using the data of Spencer et al. (12)

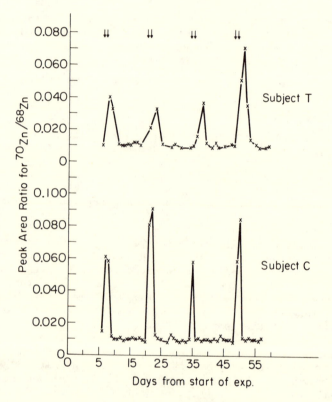

Figure 2.   Fecal excretion of ^{70}*Zn following ingestion of six equal doses over a 2-day period. Arrows indicate the 2 days of isotope administration for each diet period. For details of metabolic study, see Ref. 11.*

neglecting it would potentially underestimate true fractional
absorption by a maximum of 2%.

　　b.   Stable vs. Radioisotope Methodology.   The isotope balance technique with radiotracers has been used for estimation of
dietary mineral availability.   Thus, it is desirable to establish
by this approach, the correspondence of the stable isotope method
with its radiotracer counterpart.   We have conducted such a comparison in rats whose dietary phytate and zinc molar ratios were
varied in order to cover a range of zinc absorption (13) and
were given both ^{65}Zn and ^{70}Zn in a single intubation.   A summary
of the data is given in Table IV (13), and shows that, within the
error limits of the measurements, identical results are obtained
from the two methods.   Corresponding measurement pairs (radio and
stable isotope measurements) were found to be highly correlated
(R^2 = 0.91).

　　c.   Test of Theoretical Prediction of σ_F/F.   Although the
principles on which the prediction of σ_F/F is based have been
well established, it would be desirable to confirm their appli-
cability in an actual series of studies.   This problem can be
examined in reference to a recent study conducted in our laboratories and in which nine young adult subjects consumed simultaneously a dose of zinc chloride enriched with ^{70}Zn and a meal containing chicken intrinsically labeled with ^{68}Zn (6).   In this
experiment three metabolic periods were included and the results
are summarized in Table V and compared with the theoretical predictions discussed earlier (Tables IIIA and IIIB).   Although the
conditions used in this particular experiment are not identical
to those assumed earlier for the calculation of variability they
are sufficiently similar to make the comparisons meaningful.   The
observed variability for the fractional absorption of ^{70}Zn (Table
V) is in the range 6 - 13%, and 6 - 11% for ^{68}Zn, somewhat larger
than that predicted from consideration of measurement precision
alone (5 - 6%).   However, the variability of the ratio of absorption of ^{68}Zn to ^{70}Zn within the group of subjects is in the range
of 3 - 8% and this compares favorably with the measurement prediction of 5 - 6%.   It appears, therefore, that the predictions
of variability of measurement methodology are experimentally
validated in this double isotope experiment.

　　Methodology of Biological Labeling.   The extent of the
exchangeability of dietary zinc pools, analogous to the problem
for iron (14), must be examined in order to further explore the
application of stable isotope methodology in studies of zinc
absorption.   To accomplish this, human foods labeled appropriately
with stable isotopes of zinc must be obtained.   Although the
details of stable isotope labeling requirements vary for different foods (animal and plant), the purpose of such studies is to
achieve as high an enrichment of the zinc, or other element in
the food, as possible with a reasonable overall cost of the

isotope. Existing methods used for labeling of foods with radioisotopes are not adequate from the point of view of the required
conservation and high enrichment with stable isotopes. Therefore
we have begun a multidisciplinary effort, in collaboration with
various investigators (6, 7, 15), designed to develop and establish procedures for labeling plant and animal food with stable
isotopes of minerals. Two examples of our research are given
here:
  a. Labeling of an Animal Food Model. Overall utilization
of dietary zinc in animals is expected to be low. Thus, the
dressed meat from a broiler chicken contains approximately 8 mg
Zn. However, a total of about 2.8 kg feed are consumed during
the rearing of the chicken (6, 7), with a dietary Zn content of
40 ppm. Thus, the overall retention of dietary zinc approximates
10%. We have labeled chicken meat with ^{68}Zn, either by following
a frequent gavage schedule during the 42-day growth period (7)
or by mixing ^{68}Zn with the broiler starter ration that is fed
during the growing period (6). We observed that about 2% and 4%
of the label was incorporated into the meat, with ^{68}Zn mass
excess of 158% and 383% in the two studies, respectively. The
meat from a typical broiler provides, therefore, about 5.2 mg
total zinc per each 300 g serving of which 2.2 mg is in the form
of excess ^{68}Zn (6). The enrichment of zinc that can be observed
determines the specifics of the design of a human study in which
labeled chicken meat is to be studied. This we have considered
previously (7). Also, these findings establish the costs of isotopes for such studies, a factor that must be considered in relation to human studies of zinc bioavailability. Our experience
thus far indicates that although the present methodology of labeling permits use of such foods in human studies, further improve-
ments in efficiency of utilization of the administered dose are
very much desirable.

TABLE IV

Comparative Data on Absorption of Zinc Measured
with ^{65}Zn and Stable Isotope Methodology[1]

Treatment Group	Phytate: Zn Molar Ratio	% Absorption Measured with ^{65}Zn	% Absorption Measured with ^{70}Zn
A	0	43 ± 6	47 ± 7
B	6.25	46 ± 7	43 ± 7
C	12.5	36 ± 7	35 ± 7
D	25	33 ± 5	36 ± 5
E	50	31 ± 5	35 ± 5

[1]Mean of 5-8 animals ± 1 SEM; see ref. (13) for details

b. Labeling of a Plant Food Model. Use of stable isotopes for labeling plant foods such as wheat and soybeans requires development of non-cycling hydroponic systems (15). In a pre-liminary study (15) using ^{65}Zn as tracer, it was found that 27% and 26% of the applied dose could be recovered in the seed for soybeans and wheat respectively. Much of the remainder of unin-corporated isotope is recoverable from the culture solution. Thus, a relatively efficient means of labeling plant seeds with stable isotopes can be devised and we are in the process of testing this prediction, using ^{68}Zn and ^{70}Zn as isotopes.

Further research is needed if efficient labeling of both animal and plant food models with stable isotopes of zinc is to be achieved. This represents an important challenge in the de-velopment of stable isotope methodology for measurement of dietary zinc availability in human studies.

TABLE V

Comparisons of Fractional Zinc Absorption from an Extrinsic (^{70}Zn) and an Intrinsic (^{68}Zn) Tag[*]

Period	(1) Absorption of ^{70}Zn	(2) Absorption of ^{68}Zn	Ratio (1)/(2)
1	.46 + .06	.57 + .06	.79 + .06
2	.46 + .06	.57 + .06	.79 + .05
3	.66 + .04	.72 + .04	.91 + .03

[*] See ref. (6) for details. Data are mean + 1 SEM for 9-10 subjects.

Some Applications

Having discussed above some features of the conceptual and analytical framework concerned with stable isotope approaches for the study of zinc absorption, we present in this section summaries of a number of experiments involving application of these concepts and methods of stable isotope approaches for examination of zinc availability from human foods.

To explore the quantitative effect of various diet factors on the absorption of zinc from foods and diets, we have generally used an extrinsic tag of a stable isotope of zinc (^{70}Zn). It is assumed that this tag mixes completely with the food Zn prior to its absorption and, thus, its absorption reflects the availability of zinc in the total diet (11, 16). The validity of this assump-tion needs to be further explored (6). Furthermore, we have been concerned with the question of the dietary availability of zinc when plant foods, such as various soybean preparations, serve as the major or sole source of protein intake. For example, the results summarized in Table VI indicate that the absorption of an

"extrinsic tag" of zinc is essentially the same whether the
source of protein is either a well-processed isolated soy protein
(Supro-620, Ralston Purina Company), milk, or an isonitrogenous
mixture of these two proteins (11, 16). Additional data gathered
in other studies comparing absorption of zinc from an isolated
soy protein, soy concentrate, and animal proteins (milk and beef)
are presented in Table VII. From these findings it is reasonable
to conclude that significant intakes of high quality soy products
do not impair absorption of an "extrinsic tag" of stable isotope
of zinc when compared with proteins of animal origin. However,
because the extrinsic tag was given as a solution of inorganic
zinc it must be recognized that the conclusion may not be valid
with regard to the absorption of zinc native to the proteins of
the foods that are part of diets in which soy proteins are also
present. Furthermore, the results given in Tables VI and VII
were obtained during relatively brief experimental dietary periods of 10-14 days. Thus, it would be of interest to evaluate
the absorption of the "extrinsic zinc tag" during a more prolonged period of intake of high-quality soy product. Therefore,
we have also conducted measurements of zinc absorption at
periodic intervals within subjects who participated in a long-
term metabolic study in which the sole source of protein intake
was from a good quality soy protein concentrate (Stapro 3200,
A. E. Staley Manufacturing Co.). These results are given in
Table VIII and they reveal that the level of absorption of zinc
during the initial two weeks was maintained throughout a 3-month
period with the constant diet.

TABLE VI
Determination of Zinc Absorption in Young Men
Using ^{70}Zn and a Fecal Monitoring Method[a]

	Diet		
Total Zinc Intake	Supro 620	Supro 620/Milk	Milk
(mg/day)			
Zinc from protein	3.3	4.9	6.5
Zinc from $ZnCl_2$	12	10.5	9
Zinc absorption(%)[b]	34 ± 4	41 ± 7	41 ± 4

[a]Diets were based on milk, an isolated soy protein (Supro 620)
or a 50:50 mixture of the two sources. For more details see
(11,16).
[b]Mean ± SEM for five subjects. No significant difference
among diets.

Finally, we have completed a study designed to evaluate and
compare absorption of zinc from an "extrinsic tag" of ^{70}ZnCl$_2$
with that from an "intrinsic tag" given in the form of ^{68}Zn-
labeled chicken meat (6). These tracers were given simultaneously to young men studied during each of three experimental diet

periods: During Period 1 the protein source was chicken and Zn
intake about 10-11 mg/day. For Period 2 the protein source was
an isonitrogenous mixture of chicken and isolated soy protein
with Zn intake also being 10-11 mg/day. During Period 3 the
protein source was chicken but zinc intake was 7 mg/day. As
summarized in Table V, fractional absorption (mean $\pm$ 1 SEM for
9-10 subjects) of the extrinsic tag was .46 $\pm$ .06, .46 $\pm$ .06, and
.66 $\pm$ .04 for the three diet periods, respectively. The comparable values for the intrinsic ^{68}Zn (labeled chicken) were .57 $\pm$
.06, .57 $\pm$ .06, and .72 $\pm$ .04. There was a highly significant
correlation (R^2 = 0.86) between the fractional absorption from
the two labels, but the value for the intrinsic tag was slightly,
though significantly, higher (p < .02) than that for the extrinsic ^{70}Zn tag. Based on this initial experiment it appears that
the extrinsic tag can serve as a reliable index of absorption of
the intrinsic zinc of chicken but there is a small quantitative
discrepancy between the absolute value for absorption of the
intrinsic and extrinsic labels. The possible nutrition and metabolic significance of this difference deserves further study.

TABLE VII
Absorption of Zinc (^{70}Zn) from Diets Based on a Soy
Isolate (Supro-620) or a Soy Concentrate (Stapro 3200)
Given to Young Men (16)

Experiment	Diet (Protein Source)	Zinc Absorption (%)
1	Isolated Soy Protein	34 $\pm$ 4
	Isolated Soy Protein: Milk Combination	41 $\pm$ 7
	Milk	41 $\pm$ 4
2	Isolated Soy Protein	30 $\pm$ 3
	Beef	41 $\pm$ 4
3	Soy Protein Concentrate (Stapro 3200)	26 $\pm$ 4
	Milk	29 $\pm$ 4

Conclusions

Stable isotope methodology has recently become sufficiently
developed to permit relatively precise measurement of zinc absorption in human subjects. The methodology using Radiochemical
Neutron Activation Analysis has been validated and is now being
applied to the study of the effects of various dietary and host
factors on the availability of dietary zinc. However, additional
analytical developments and refinements will be necessary before
a maximum utilization of this safe and non-invasive method is

made in studies of zinc nutrition and metabolism in man. With further refinement in the overall measurement methodology, the opportunity will arise to conduct various investigations into the practical aspects of zinc nutriture in humans that cannot be carried out at this time. Finally, considerable additional research is needed in order to make available a variety of foods labeled with stable isotopes of zinc for use in studies of exchangeability of dietary pools and the availability of native food zinc. Current research advances lead to the encouraging conclusion that it will be possible to achieve an accurate means, based on non-invasive approaches involving stable isotopes, for exploring many quantitative aspects of dietary zinc availability in humans under various pathophysiological states.

TABLE VIII

Zinc Absorption, Determined with an Extrinsic ^{70}Zn, at Intervals During a Long-Term Metabolic Study with a Soy Protein Concentrate (Stapro 3200), in Young Adult Men (16)

Period and Day of Study	Zinc Absorption (%)
Initial (egg)	31 + 5[1]
Soy Concentrate[2]	
Day 12	22 + 4
Day 24	19 + 4
Day 36	23 + 3
Day 48	22 + 5
Day 60	32 + 5
Day 72	25 + 3
Day 84	19 + 3

[1]Mean + SEM for 6 young adult subjects.

[2]Soy concentrate was sole source of protein intake (0.8 g/kg/day).

Acknowledgements

The work reported here has been supported in part by grants from NSF (#7919112-PFR), USDA (Competitive Research Grants Office, CSRS, S&E, 59-2253-1-1-769-0), Ralston Purina Company (St. Louis, MO), and A. E. Staley Manufacturing Company (Decatur, IL), for which the authors are grateful.

Literature Cited

1. Sandstead, H. H. Am. J. Clin. Nutr. 1973, 26, 1251–60.
2. Prasad, A. S., Ed.; "Trace Elements in Human Health and Disease"; Vol. I; Academic Press: New York, 1976; p 470.
3. Solomons, N. W. Am. J. Clin. Nutr. 1982, 35(5):1048–1075.
4. Aamodt, R. L.; Rumble, W. F.; Johnston, G. S.; Foster, D.; Henkin, R. I. Am. J. Clin. Nutr. 1979, 32, 559–69.
5. Weigand, E.; Kirchgessner, M. J. Nutr. 1980, 110, 469–80.
6. Janghorbani, M.; Istfan, N. W.; Pagounes, J. O.; Steinke, F. H.; Young, V. R. Am. J. Clin. Nutr. 1982, in press.
7. Janghorbani, M.; Ting, B. T. G.; Young, V. R.; Steinke, F. H. Br. J. Nutr. 1981, 46, 395–402.
8. Janghorbani, M.; Ting, B. T. G.; Young, V. R. Clin. Chim. Acta 1980, 108, 9–24.
9. Prasad, A. S., Ed.; "Clinical, Biochemical, and Nutritional Aspects of Trace Elements", Alan R. Liss, Inc.: New York, 1982, in press, Chapter 24.
10. DeSoete, D.; Gijbels, R.; Hoste, J. "Neutron Activation Analysis"; Wiley-Interscience: New York, 1972; p 836.
11. Solomons, N. W.; Janghorbani, M.; Ting, B. T. G.; Steinke, F. H.; Christensen, M. J.; Bijlani, R.; Istfan, N. W.; Young, V. R. J. Nutr. 1982, in press.
12. Spencer, H.; Rosoff, B.; Feldstein, A.; Cohn, S. H.; Gusmano, E. Rad. Res. 1965, 24, 432–35.
13. Lo, G. S.; Steinke, F. H.; Ting, B. T. G.; Janghorbani, M.; Young, V. R. J. Nutr. 1981, 111, 2236–9.
14. Hallberg, L. Ann. Rev. Nutr. 1981, 1, 123–47.
15. Weaver, C. M.; Janghorbani, M.; Young, V. R. Fed. Proc. 1982, in press.
16. Young, V. R.; Janghorbani, M. J. Plant Foods 1982, in press.

RECEIVED October 6, 1982

Zinc Absorption in Humans
Effects of Age, Sex, and Food

R. L. AAMODT and W. F. RUMBLE
National Institutes of Health, Bethesda, MD 20205

ROBERT I. HENKIN
Georgetown University Medical Center, Center for Molecular Nutrition and
Sensory Disorders, Washington, DC 20007

We have studied human Zn-69m and Zn-65 metabolism
using a detailed kinetic model and the SAAM27
computer program. Absorption of Zn-65 by 75 fasting normal volunteers was 65+11% (mean+ 1SD), range
40-86%). Absorption related to age (by decade)
decreased linearly A=69.5-0.123Y (A=absorption,
Y=age in years), correlation coefficient(r)=-0.934.
Although mean absorption by women was higher (67+
2%, mean+1SEM) than that of men (63+2%) this difference was not statistically significant (ANOVA or
t-test for comparison of means). Absorption related to age (by decade) was significantly less
(p<0.05) for men than for women (paired t-test).
Three subjects who ate breakfast before receiving
Zn-65 had significantly lower absorption (17.4%,
range 9-31%) than fasting subjects. Although human
zinc absorption decreased with age, the mechanism
of this change is unclear. Food decreases zinc
absorption but specific food effects are not well
defined.

Zinc metabolism is a dynamic process which cannot be fully
defined by any single static measurement. Since it is related in
part to absorption, zinc bioavailability can be reduced by
abnormalities in the gastrointestinal tract, in transport ligands
or substances in the intestinal contents which interfere with
zinc absorption. Bioavailability can also be affected by metabolic defects which prevent zinc uptake or utilization after
absorption and by several additional factors. Some of these
additional factors have been considered by previous investigators
and include age, sex, food and the initial conditions under which
bioavailability and metabolism were determined.

A number of processes related to absorption and metabolism
occur simultaneously following the ingestion of zinc (Figure 1).

0097-6156/83/0210-0061$06.50/0

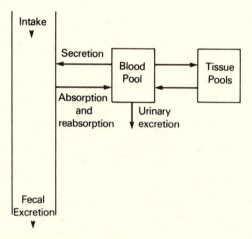

Figure 1. A simple model illustrating some of the complex interactions involved in zinc absorption and metabolism. Following absorption, a fraction of the zinc goes directly into blood, another fraction passes first through the portal circulation and then into the systemic circulation. Some zinc is resecreted into the gastrointestinal tract, thereby becoming available for reabsorption. Activity in blood can exchange with tissue pools, be excreted in urine, or secreted back into the gastrointestinal tract.

A fraction of the ingested zinc is absorbed, but the rate of absorption is not constant throughout the gastrointestinal tract(1). Some of the absorbed zinc enters the systemic circulation directly, some enters the portal circulation first, then passes into the general circulation, and some is secreted back into the gut and is then available for reabsorption.

One effective method of obtaining information about zinc metabolism is the use of radioactive tracers to study absorption, kinetics and body distribution following oral administration. Of course, as in all cases in which radionuclides are administered to human subjects, the hazard of such administration must be considered and balanced against the value of information expected from the study. The administered activity should be as low as possible consistent with obtaining adequate measurements. The studies reported here were carried out in accordance with approved protocols which followed relevant guidelines for studies of human subjects and with the informed consent of the participating subjects.

The following is a summary of results of zinc metabolism studies of patients and normal volunteers carried out over the past 10 years. Initially nearly all the absorbed zinc is in plasma, but it is rapidly taken up by the liver (Figure 2). During the first 5 to 48 hours approximately 70% of the absorbed radioactive zinc tracer goes to the liver(1,2,3). During the next 50 days liver zinc decreases while Zn-65 activity in the thigh (primarily muscle zinc) increases (3). Uptake in the thigh appears to mirror losses from the liver until after about 100 days when both have similar activity and decrease at about the same rate. Red blood cells begin to take up zinc rapidly, with activity appearing as early as 15 minutes after administration and increasing thereafter to reach maximum values 7 to 14 days later(3). Total body retention decreases rapidly at first as unabsorbed zinc is excreted, then more slowly(3). After about 100 days all body compartments lose zinc very slowly with half-times of approximately 400 days(3).

In order to understand these complex metabolic interactions more fully and to maximize the information obtained in these studies, we developed a detailed kinetic model of zinc metabolism(1,2). Modeling of the kinetic data obtained from measurements of biological tracers by compartmental analysis allows derivation of information related not only to the transient dynamic patterns of tracer movements through the system, but also information about the steady state patterns of native zinc. This approach provides data for absorption, absorption rates, transfer rates between compartments, zinc masses in the total body and individual compartments and minimum daily requirements. Data may be collected without disrupting the normal living patterns of the subjects and the difficulties and inconveniences of metabolic wards can be avoided.

The initial zinc model was based upon averaged data for 17

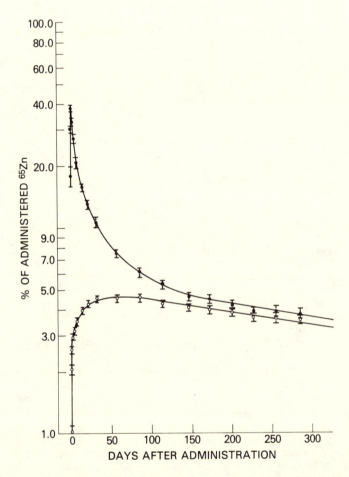

Figure 2. Percentage of [65]*Zn activity present in liver (○) and thigh (●) areas of normal subjects measured by external γ-ray detectors during a 350-day period after oral administration. Activity in the liver area increased rapidly during the period between 5 and 48 h to approximately 70% of the absorbed activity (49% of administered activity), then decreased. Activity in the thigh area increased more slowly, reaching maximum values by about 50 days, then decreasing. After approximately 100 days after administration both liver and thigh areas had similar activities and loss rates.*

patients with defects of taste and smell given Zn-69m both orally
and intravenously two weeks apart(2). These data were analyzed
using the SAAM27 computer program(4) and a model for zinc
metabolism(1) was derived (Figure 3). The compartmental model
used all the kinetic data obtained from Zn-69m activity in
plasma, red blood cells, liver and thigh as well as stable zinc
parameters, including: dietary intake, serum levels and urinary
concentration. These values were used to obtain the simplest set
of mathematical relationships that would satisfy all of the data
characteristics for each measurement time in the study while
remaining consistent with accepted concepts of zinc metabolism.

Despite the short half-time of Zn-69m which limited the data
collection period such that the model could only analyze the
rapid processes of zinc metabolism (approximately 10% of total
body zinc), a number of fundamental steady state parameters were
derived as shown in Table I.

TABLE I.

FUNDAMENTAL STEADY STATE ZINC PARAMETERS
DERIVED FROM THE Zn-69m MODEL.

PARAMETER	MEASURED VALUE	CALCULATED VALUE	NORMAL VALUE
Plasma zinc concentration (ug/ml)	0.90	1.07	1.0
Daily plasma zinc intake (mg/day)	--	2.7	--
Red cell zinc concentration (ug/ml)	--	21.6	14.4
Blood zinc in cells (%)	--	92	75-88
Liver zinc concentration (ug/ml)	--	69.5	55-76
Urinary zinc excretion (mg/day)	0.72	0.72	0.3-0.6
Daily zinc intake (mg/day)	--	7.3	--
Absorption (%)	--	37	--

The model also considered the absorption process (Figure 4)
as occurring in a series of compartments with outputs to plasma,
red blood cells and liver(1). The model thus allowed the definition of absorption as a kinetic process and the separate consideration of different aspects of this process (such as the
fraction of absorbed zinc which enters directly into plasma and
changes in absorption rate as a function of time) from which
absorption in various parts of the gut could be obtained (1).

Because the short half-time of Zn-69m limited these studies
to the early phases of zinc metabolism, they were continued with
the longer lived nuclide Zn-65(3). This nuclide allowed the
development of an extended model of zinc metabolism(5) which
included data for both rapid and slower processes. This model
was discussed in detail in references 5 and 6 and will be devel-

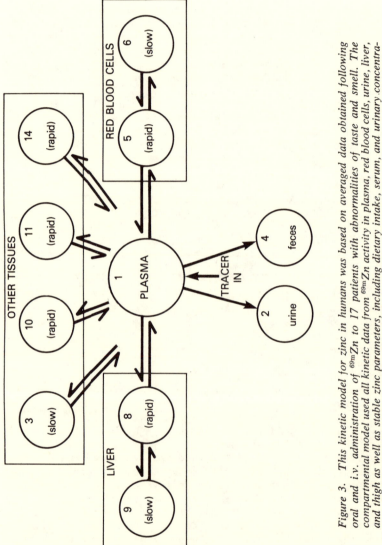

Figure 3.   This kinetic model for zinc in humans was based on averaged data obtained following oral and i.v. administration of ⁶⁹ᵐZn to 17 patients with abnormalities of taste and smell. The compartmental model used all kinetic data from ⁶⁹ᵐZn activity in plasma, red blood cells, urine, liver, and thigh as well as stable zinc parameters, including dietary intake, serum, and urinary concentration. The SAAM27 computer program was used to obtain the simplest set of mathematical relationships that would satisfy the data characteristics for each measurement time in the study and remain consistent with accepted concepts of zinc metabolism. Although the short physical half-life of ⁶⁹ᵐZn limited the data collection period, this model allowed for analysis of the rapid phases of zinc metabolism (about 10% of total body zinc) and derivation of a number of fundamental steady state parameters (see Table I).

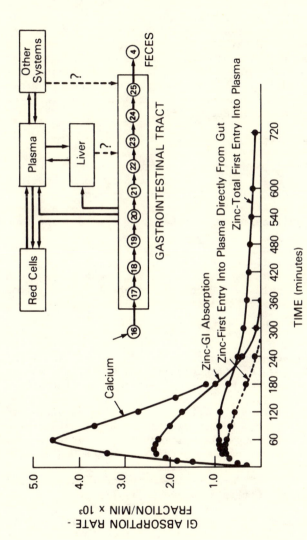

Figure 4. The ⁶⁹ᵐZn model expanded to illustrate the compartments used to analyze the process of zinc absorption. Compartment 16 represents stomach and 17–25 represent the intestine as a set of compartments connected in series with outputs to red blood cells, plasma, and liver. The model allows the definition of absorption as a kinetic process and the separate consideration of various aspects of this process. The model then could be used to define the kinetics, not only of total zinc absorption, but also of zinc entry directly from the gut into plasma, entry into plasma following uptake by liver and red blood cells, and absorption rates at various times after administration. Absorption in various parts of the gut can be derived from these kinetic data.

oped further in connection with studies of zinc pathology in the next paper(7).

The techniques used to obtain these data and to define the extended model of zinc metabolism are important to define changes in zinc metabolism in both normal and pathological states and are particularly useful as an aid in definition of zinc deficiency. Human zinc deficiency has been difficult to identify because its symptoms are complex and variable (6,8-11) ranging from impairments of taste and smell to skin rash and dwarfism. There is presently no adequate clinical test for the diagnosis of zinc deficiency in humans, which further complicates its identification. Zinc content in serum, plasma, red blood cells,urine, saliva or hair have proven inadequate for specific diagnosis of individual cases of zinc deficiency(6,8-12).

The purposes of this paper are:

1. To present the various methods used to define and estimate absorption in order to detail differences in methodology which may create confusion when results are compared.
2. To discuss data related to zinc absorption by normal human subjects, specifically the effects of age, sex and food intake.
3. To indicate processes other than absorption which also affect zinc bioavailability.
4. To describe how computer modeling of zinc metabolism can be used to elucidate the complex metabolic interactions related to zinc bioavailability.

ABSORPTION

Gastrointestinal absorption may be defined as the process by which substances which enter the digestive tract are transported by either active or passive mechanisms across the intestinal epithelium into blood or lymph.

While adequate intestinal absorption is clearly required to maintain zinc sufficiency, a number of different methods are presently used for its measurement. Intestinal absorption has been measured by clinical balance studies, tracers (using both stable and radioactive isotopes) and serum levels after ingestion of pharmacologic doses of stable zinc compounds. Each can be useful, but each also has limitations and it is important to note that different methods, in reality, measure different aspects of absorption. While results are often reported as absorption, specific nomenclature has been developed to specify the results from each method of measurement(13,14).

NET ABSORPTION applies only to chemical balance studies(14) and is defined as the difference between intake and fecal excretion during a balance period. Because net absorption assumes, at least tacitly, that all fecal excretion consists of material which was not absorbed after entering the digestive tract, it will be correct only for substances such as potassium and sodium which are primarily excreted by non-fecal routes. Zinc, on the contrary, is excreted primarily in feces and endogenous excretion can be a significant fraction of fecal excretion. Net absorption has 'the peculiar property that when intake is reduced or endogenous secretion increased it can become negative while continued absorption can be clearly demonstrated by tracer techniques.

APPARENT ABSORPTION is another commonly determined measurement of absorption. It is equivalent to net absorption for a single oral administration of stable or radioactive tracer. Apparent absorption can be defined as the difference between administered activity and fecal excretion during a period long enough for excretion of unabsorbed activity from the gastrointestinal tract to occur. While this quantity cannot become negative, since only administered tracer is measured, it understates absorption by not excluding from fecal excretion the fraction of administered activity which was absorbed and resecreted into the gastrointestinal tract during the study period.

TRUE ABSORPTION accounts for all absorbed activity including that which is secreted back into the gut. True absorption meets the definition given previously and accounts for all material which crosses the intestinal epithelium into blood or lymph. True absorption, however, cannot be easily determined except in those rare cases in which endogenous secretion into the gut is minimal or can be determined separately from fecal excretion of unabsorbed substance.

FRACTIONAL ABSORPTION is the ratio of activity in total body, organs or tissues after oral administration to that after parenteral administration. The method for measurement of fractional absorption assumes that once a substance enters the body it is metabolized similarly regardless of the route of administration.

Another method, useful in some applications, is measurement of serum or urinary levels of a given substance following oral administration of pharmacologic doses of test substance. These techniques are similar in concept to tolerance tests such as those for glucose, folic acid or vitamin B-12 and like those may provide clues for determination of absorptive abnormalities. These techniques do not, however, provide a means to assess absorption since they involve so many uncontrolled variables. For zinc, several such studies have recently been conducted (15-20). We have previously shown that levels of tracer zinc in plasma, red blood cells, whole blood(21) or urine(22) are closely correlated with absorption when measured under specific controlled conditions. These conditions include measurement of subjects fasted overnight and oral administration of carrier free tracer. The relationship between plasma Zn-69m activity and absorption estimated from total body retention of Zn-69m was determined using previously described techniques(1) in samples taken at various times after administration. Using this method it was demonstrated that activity in plasma taken two to five hours after ingestion correlated closely with fractional absorption (21). Similarly, concentration in washed red blood cells taken 1 to 4 days after ingestion or whole blood taken in either of these periods correlated closely with absorption (Table II). Using a similar technique Malokhia et al.(23) later determined absorption using deconvolution analysis of serum activity over an eight hour period following separate oral and intravenous administration of Zn-69m.

TABLE II.
CORRELATION COEFFICIENT
(% absorption of Zn-69m measured by total body counting versus % absorption measured using blood fraction values)
number of patients = 17

TIME	WHOLE BLOOD	PLASMA	CELLS
20 min.	0.51		
30 min.	0.59		
45 min.	0.71		
60 min.	0.78		
1.5 hr.	0.86		
2.0 hr.	0.85	0.82	
3.0 hr.	0.86	0.83	
4.0 hr.	0.80	0.73	
5.0 hr.	0.73	0.63	
1 day	0.80	0.83	0.77
2 day	0.81	0.72	0.73
3 day	0.74	0.53	0.75
4 day	0.70	--	0.76

While absorption and concentration in blood and urine appear to be closely correlated under these conditions, a number of factors such as food or specific food components (ie: fiber, protein or minerals) and history of zinc nutriture may affect plasma or urinary levels, especially in individual samples taken at specific times after ingestion. Changes in zinc levels in whole blood, plasma or serum over time are the result of several competing processes as illustrated in Figure 1. These include:
1. Pre-existing or baseline blood zinc levels.
2. The magnitude and rates of zinc absorption, reabsorption and secretion.
3. The pattern of zinc absorption as a function of time.
4. The rate of entry of zinc into blood.

5. Rates of loss of zinc from blood to tissues.
6. Rate of loss of zinc from blood by excretion.

Factors which change these processes may produce effects on zinc levels in blood or blood components which are not easily predictable. Food may affect intestinal transit time, bind zinc or decrease exposure of zinc to the mucosal surface, effects which may alter total absorption, the time of maximal absorption or absorption rates(24). Studies which measure the difference between fasting baseline levels of stable zinc in serum and those after administration of a zinc load with a test meal may underestimate the degree of response since food alone (without added zinc) has been reported to lower serum zinc levels (16,17, 18,25-28). The effect of changes in absorption pattern on serum, plasma or blood levels is shown in Figure 5. This illustration simulates blood concentration over time for three situations in which total absorption of a substance was held constant as the time of peak concentration was varied from one to two to three and one half hours after ingestion. This figure illustrates that minor changes in absorption pattern can produce large changes in blood concentration values measured at fixed times after administration.

Factors other than food can also affect zinc absorption and plasma zinc levels. We have recently shown that zinc absorption in patients with defects of taste and smell taking 110 mg/day of Zn^{++} (100 mg/day as ZnSO)(given daily as four 25 mg doses, the remaining 10 mg of dietary origin) decreased to 21% of that observed with 10 mg/day of dietary origin while daily absorption increased from 7.8 to 11.5 mg/day(5). Finally, zinc has been shown to follow a circadian pattern with high levels noted in serum samples taken between 10:00 A.M. and 10:00 P.M. and low levels between 2:00 A.M. and 7:00 A.M(29).

These factors emphasize the need for caution in interpreting results of absorption studies or studies which relate blood or urinary levels to absorption. It is particularly important that absorption measurements be designed systematically to exclude or control variables such as food, phytate, fiber and all other factors which separately or in combination could confound interpretation of the results of even well executed studies.

STUDIES OF ABSORPTION BY NORMAL SUBJECTS

While there have been a number of studies of zinc metabolism in various patient populations using Zn-69m and Zn-65 tracers(1,3,5,30-34), few of these have included data for normal subjects(23,33,34,35) and then only for limited numbers of normals. We have been aware for some time that data about zinc metabolism in normal subjects was required in order to interpret data from patients and that by and large these data did not exist.

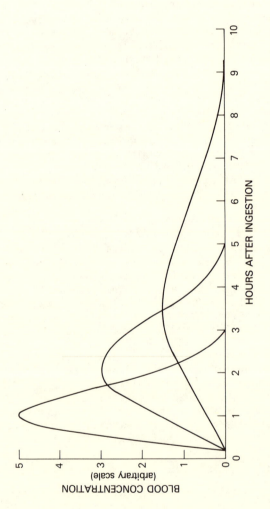

Figure 5. A simulation of the pattern of blood concentration over time for three situations in which total absorption of a given substance remained constant as the time of peak concentration was varied from 1 to 2 to 3½ h after ingestion. This figure illustrates that minor variations in absorption pattern can produce large changes in blood concentration values measured at fixed times after administration.

During the past three years we have studied absorption and metabolism of Zn-65 in normal subjects to compare with data we have obtained in patients suspected of having abnormal zinc metabolism. These subjects received 10 μCi of Zn-65 as carrier free zinc chloride orally after an overnight fast and absorption and metabolism were measured using previously published techniques(4,36). These studies were designed to minimize uncontrolled variables in order to establish a basis for rigorous and systematic comparisons between normal subjects and patients with abnormalities of zinc metabolism. Subjects in these studies were recruited from the community and had a wide variety of backgrounds. None of these subjects was taking any prescription medications nor had any symptoms of acute or chronic diseases.

While zinc is generally consumed with or in food, the tracer was given after an overnight fast and food was not allowed for at least five hours after administration. Administration of the tracer with food would have introduced at least one and probably several new uncontrollable variables into the study. Sandstrom et al.(37,38) have shown that Zn-65 absorption can be variably increased or decreased by such dietary factors as white versus whole grain bread and levels of zinc, calcium and protein.

Data for intestinal absorption of Zn-65 in 75 normal volunteers (31 women, 44 men, aged 18-84 years) have recently been published(36) and more recently the effects of age and sex on absorption have been evaluated. Three subjects who had eaten breakfast prior to receiving oral Zn-65 were also studied and their data compared with that from fasting subjects.

Fractional absorption was determined as the mean of ratios of total body retention values for the 75 normal volunteers measured 7,14 and 21 days after oral administration of Zn-65 to similar measurements following intravenous administration to two patients as described previously(35).

Variations among individual absorption values calculated from 7,14 and 21 day retention values and the mean of all three are shown in Figure 6. For each subject absorption values were similar on each of the three measurement days. Values measured on day 7 were slightly less than the mean (-0.79%) while those measured on day 14 were slightly higher (0.81%) and those measured on day 21 essentially equal to the mean. These differences were small and evenly distributed indicating that fecal excretion was more than 99% complete in these subjects by day 7, and the similar variability of the data for each measurement time strongly suggests that most differences were due to measurement variability rather than to delayed fecal excretion of unabsorbed activity.

The absorption results for the 75 fasting normal subjects are shown in Figure 7. Mean absorption of carrier free Zn-65 was 65+11% (mean+1SD) with a range of 40 to 86%. These values had a higher mean and a narrower range than those for patients with

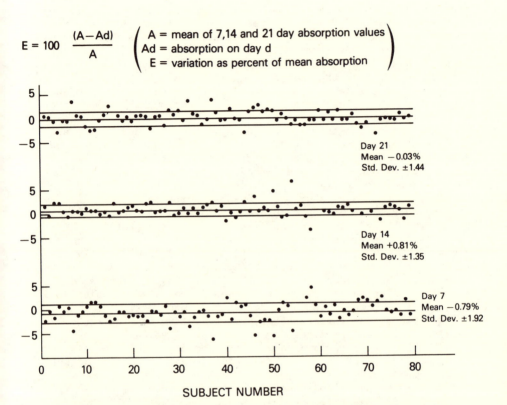

Figure 6. *Variability of intestinal zinc absorption calculated from retention measured on days 7, 14, and 21 after oral administration of ⁶⁵Zn to fasting normal volunteers. All measurements were in good agreement: values measured on day 7 were slightly less than the mean (−0.79%), those on day 14 slightly more than the mean (+0.81%), and those measured on day 21 essentially equal to the mean. These small and evenly distributed differences indicate that fecal excretion was more than 99% complete by day 7. The similar variability obtained for each measurement period strongly suggests that most differences were related to measurement variability rather than to delayed fecal excretion.*

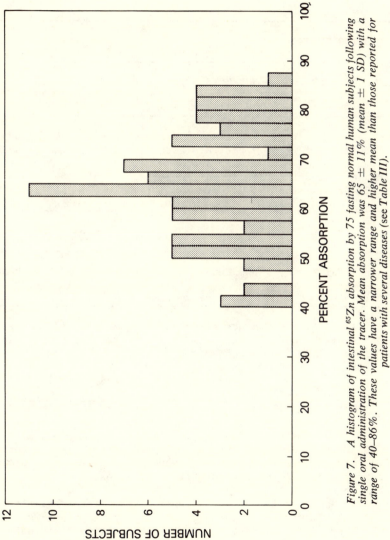

Figure 7. A histogram of intestinal [65]Zn absorption by 75 fasting normal human subjects following single oral administration of the tracer. Mean absorption was 65 ± 11% (mean ± 1 SD) with a range of 40–86%. These values have a narrower range and higher mean than those reported for patients with several diseases (see Table III).

several diseases previously reported by ourselves and by others
(Table III).

TABLE III.

EFFECTS OF FOOD ON INTESTINAL ZINC ABSORPTION

I. FASTING SUBJECTS

INVESTIGATOR		SUBJECTS	NUCLIDE	MEAN % ABSORP.	RANGE
AAMODT et al.	1981	75 normal volunteers	Zn-65	65	40-86
HAWKINS et al.	1976	9 patients with skin diseases	Zn-65	74	34-100
LOMBECK et al.	1975	5 control subjects	Zn-65	66	58-77
MOLOKHIA et al.	1980	5 volunteers	Zn-69m	61	41-79

II. FED SUBJECTS

INVESTIGATOR		SUBJECTS	NUCLIDE	MEAN % ABSORP.	RANGE
AAMODT et al.	1981	3 normal volunteers	Zn-65	17	9-31
ARVIDSSON et al.	1978	11 subjects	Zn-65	27 [*]	18-39
MOLOKHIA et al.	1980	3 subjects	Zn-69m	<20 [**]	--
SANDSTROM et al.	1980	66 subjects	Zn-65	10-38 [**]	6-52
	1980	35 subjects	Zn-65	14-36 [**]	11-47

[*] Mean not given [**] Range of mean values for various conditions

AGE EFFECTS Mean absorption decreased in a linear monotonic
manner with increasing age as shown in Figure 8. The plotted
curve is a linear least squares fit to the experimental points
which are shown with a range of ± 1SEM. The decrease in absorption with age could be expressed by the relationship
A=69.5-0.123Y, where A is percent absorption and Y is age in
years. These data indicate that zinc absorption decreases by
about a tenth of a percent per year over the age range studied
and that the decrease was highly correlated with age (correlation
coefficient -0.934, p<0.001).

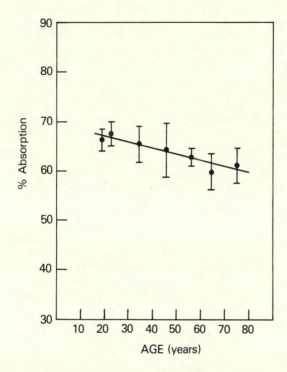

Figure 8. The relationship between age and absorption for the 75 normal subjects. Mean absorption decreased in a linear monotonic manner with increasing age. The plotted curve is a linear least squares fit to the experimental points which are mean values with a range of ± 1 SEM. The decrease in absorption with age can be expressed by the relationship $A = 69.5 - 0.123 Y$ (A = % absorption, Y = age in years).

SEX AND AGE EFFECTS Absorption was also correlated with   age
for both men and women when data for each was plotted separately
by age group (Figure 9). As indicated in this figure, absorption
was greater for young women than for young men, but absorption
decreased more rapidly with age in women than in men. The
relationship between absorption and age for women can be expressed as A=78.4-0.254Y with a correlation coefficient of -0.887
while that for men was A=67.7-0.128Y with a correlation coefficient of -0.859. Women absorbed more zinc than men in all but
one age category and these differences were significant (p<0.05)
when absorption was related to age (by decade) using a paired-t
test. Zinc absorption by women, while higher (67±2%, mean ±
1SEM) than that for men (63±2%), was not significantly different
by t-test for comparison of means or by analysis of variance with
age excluded (ANOVA).

FOOD EFFECTS The effects of food on zinc absorption are
important and particularly pertinent to the topic of this symposium. Generally, food appears to interfere with zinc absorption. Three subjects who had eaten breakfast one to three hours
prior to receiving Zn-65 absorbed a mean of 17% of the administered activity with values ranging from 9 to 31%. Absorption by
these three non-fasting subjects averaged about one quarter that
of the 75 fasting subjects studied in the same manner.
     Several investigators have studied zinc absorption in either
fasting or fed subjects, although few have studied both. Generally tracer zinc was less well absorbed when given with food than
when given to fasting subjects as shown in Table III. As was the
case for our limited studies, these data indicate that absorption
of zinc by fed subjects was about a quarter that by fasting
subjects who averaged about 66%.
     The complexity of food effects on zinc absorption is il-
lustrated by the studies of Sandstrom et al.(25,37,38) in which
phytate, protein, calcium, zinc and other factors appear to have
variable effects on zinc absorption. Although it appears certain
that food interferes with zinc absorption, the effects of individual food substances are unknown and difficult to predict.
Regardless of individual effects, zinc administration with food
will complicate efforts to measure the effects of other variables
on intestinal absorption.

SUMMARY AND CONCLUSIONS

     The intent of this paper has been to discuss some approaches
to the determination of zinc bioavailability both in terms of
absorption measurements and in terms of factors other than
absorption which affect zinc metabolism. We also presented the
results of zinc absorption studies in normal subjects and the
methodology which was developed to interpret these studies. Zinc
metabolism is a series of complex dynamic processes which cannot

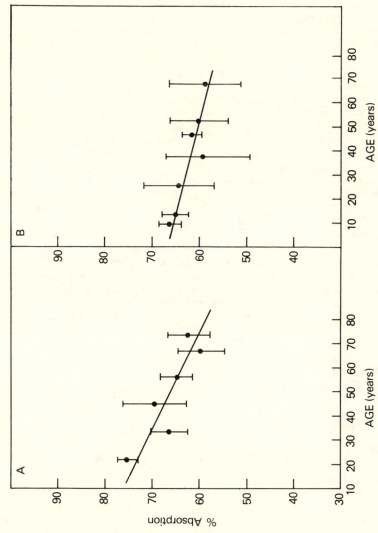

Figure 9. The relationship between age and absorption for women (A) and men (B). Absorption was greater for young women than for young men but decreased more rapidly with age for women (women: A = 78.4 − 0.254Y; men: A = 67.7 − 0.128Y; A = % absorption, Y = age in years).

be fully defined by single static measurements such as zinc concentration in urine or serum measured either directly or following administration of a load of exogenous zinc. In order to evaluate these complex processes, we have developed a detailed multi-compartmental model which allows simultaneous evaluation of events occurring in a number of body organs and tissues and determination of the complex interactions between tissues which are otherwise difficult or impossible to obtain.

We have studied zinc absorption in 78 normal volunteers (75 fasting, 3 fed) using the tracer Zn-65. Intestinal absorption by fasting normal subjects ranged from 40 to 86% with a mean of 65+11% (mean+1SD). These studies are the first to demonstrate that zinc absorption in humans decreased with age. While absorption by men was significantly different from that by women when related to age (by decade) using paired t-test, neither t-test for comparison of means nor analysis of variance with age excluded indicated significant differences. We conclude that the evidence is not sufficient to establish the existence of such differences. Food dramatically decreased zinc absorption to a level about one quarter that measured in fasting subjects based upon data for the three subjects in our studies who had had breakfast prior to taking tracer and that from published values for other studies of fasting and fed subjects.

These are the first studies of zinc absorption reported in a large group of normal subjects. Results of these studies are important as a basis for comparison with data from patients with various diseases who may exhibit abnormalities of zinc absorption. Zinc absorption has been shown to decrease with age in normal humans, but whether this is related to decreased anabolism with reduced demand for zinc or to some physiological or pathological process associated with aging is unclear. The decreased absorption of zinc taken with food is important with respect to the design of studies of zinc metabolism, in the determination of zinc bioavailability and nutritional requirements of the trace metal and in treatment of subjects who may require zinc replacement.

Literature Cited

1. Foster, D.M.; Aamodt, R.L.; Henkin, R.I.; Berman, M. *Am. J. Physiol.* 1979, 237, r340-347.
2. Aamodt, R.L.; Rumble, W.F.; Johnston, G.S.; Foster, D. *Am. J. Clin. Nutr.* 1979, 32, 559-569.
3. Aamodt, R.L.; Rumble, W.F.; Babcock, A.K.; Foster, D.M.; Henkin, R.I. *Metabolism.* 31, 326-334.
4. Berman, M.; Weiss, M.F. SAAM Manual. U.S. Government Printing Office, Washington, D.C. 1978. (DHEW Publ. No. (NIH)76-730).

5. Babcock, A.K.; Henkin, R.I.; Aamodt, R.L.; Foster, D.M.; Berman, M. Metabolism. 31, 335-347.

6. Henkin, R.I.; Aamodt, R.L.; Babcock, A.K.; Agarwal, R.D.; Shatzman, A.R. "Perception of Behavioral Chemicals"; Norris, D.M. Ed.; Elsevier/North Holland Biomedical Press:Amsterdam, 1981, pp. 229-265.

7. Henkin, R.I.; Aamodt, R.L. chapter ___ in this book.

8. Aggett, P.J.; Harries, J.T. Arch. Dis. Childhood. 1979,54, 909-917.

9. Prasad, A.S. C.R.C Crit. Rev. in Clin. Lab. Sci. 1977, 8, 1-80.

10. Walravens, P.A. West. J. Med. 1979,130, 133-142.

11. Henkin, R.I.; Apgar, J.; Cole, J.F. "Zinc" University Park Press: Baltimore, 1979.

12. Solomons, N.W. Am. J. Clin. Nutr. 1979, 32, 856-871.

13. Weigand, E.; Kirchgessner, M. J. Nutr. 1980, 110, 469-480.

12. Holdsworth, CD. "Intestinal Absorption in Man"; McColl, I. and Sladen, G.E. Eds.; 1975, Academic Press:New York, pp. 223-262.

15. Pecaud, A.; Donzel, P.; Shelling, J.L. Clin. Pharmacol. Ther. 1975, 17, 469-474.

16. Oelshlegal, F.J.; Brewer, G.J. "Zinc Metabolism: Current Aspects in Health and Disease"; Brewer, G.J.; Prasad, A.S. Eds.; Alan R. Liss Inc.: New York, 1977, pp. 299-316.

17. Solomons, N.W.; Jacob, R.A.; Pineda, O.; Viteri, F.E. J. Nutr. 1979, 109, 1519-1528.

18. Solomons, N.W.; Jacob, R.A.; Pineda, O.; Viteri, F.E. J. Lab.Clin. Med. 1979, 94, 335-343.

19. Solomons, N.W.; Jacob, R.A.; Pineda, O.; Viteri, F.E. Am. J. Clin. Nutr. 1979, 32, 2495-2497.

20. Solomons, N.W.; Jacob, R.A. Am. J.clin. Nutr. 1981, 34, 475-482.

21. Rumble, W.F.; Aamodt, R.L.; Henkin, R.I. Fed. Proc. 1977, 36, 1138 (abstract).

22. Henkin, R.I.; Aamodt, R.L. Lancet. 1975, I, 1379-1380 (letter).

23. Molokhia, M.; Sturniolo, G.; Shields, R.; Turnberg, L.A. Am. J. Clin. Nutr. 1980, 33, 881-886.

24. Sladen, G.E.; "Intestinal Absorption in Man", op. cit. pp1-49.

25. Sandstrom, B.; Arvidsson, B.; Cederblad, A.; Bjorn-Rasmussen, E. Am. J. Clin. Nutr. 1980, 33, 739-745.

26. Kasperek, K.; Schicha, H.; Siller, V.; Feinendegen, L. Strahlentherapie. 1973, 145, 229-233.

27. Davies, I.J.T.; Musa, M.; Dormandy, T.L. J. Clin. Path. 1968, 21, 359-365.

28. Harland, B.F.; Spivey-Fox, M.R.; Fry, B.R. Jr. Proc. Soc. Exp. Biol. Med. 1974, 145, 316-322.

29. Lifschitz, M.D.; Henkin, R.I. J. Appl. Physiol. 1971, 31, 88-92.

30. Spencer, H.; Vankinscott, V.; Lewin, I.; Samachson, J. J.
     Nutr. 1965, 86, 169-177.
31. Spencer, H.; Rosoff, B.; Lewin, I.; Samachson, J. "Zinc
     Metabolism";Prasad, A.S. Ed., Charles C. Thomas:Springfield,
     1966, pp. 339-361.
32. Hawkins, T.; Marks, J.M.; Plummer, V.M.; Greaves, M.W. Clin.
     Exp. Dermatol. 1976, 1, 243-252.
33. Lombeck, I.; Schnippering, H.G.; Ritzl, F.; Feinendegen,
     L.E.; Bremar, H.J. Lancet. 1975, I., 855.
34. Weissman, K.; Hoe, S.; Knudson, L.; Sorenson, S.S. Brit. J.
     Dermatol. 1979, 101, 573-579.
35. Richmond, C.R.; Furchner, J.E.; Trafton, G.A.; Langham, W.H.
     Hlth. Phys. 1962, 8, 481-489.
36. Aamodt, R.L.; Rumble, W.F.; Johnston, G.S.; Markley, E.J.;
     Henkin, R.I. Am. J. Clin. Nutr. 1981, 34, 2648-2652.
37. Arvidsson, B.; Cederblad, A.; Bjorn-Rasmussen, E.;
     Sandstrom, B. Int. J. Nucl. Med. Biol. 1978, 105,
     104-109.
38. Sandstrom, B.; Cederblad, A. Am. J. Clin. Nutr. 1980, 33,
     1778-1783.

RECEIVED October 25, 1982

A Redefinition of Zinc Deficiency

ROBERT I. HENKIN

Georgetown University Medical Center, Center for Molecular Nutrition and Sensory Disorders, Washington, DC 20007

ROGER L. AAMODT

National Institutes of Health, Bethesda, MD 20205

Based upon clinical data we have reclassified zinc deficiency using traditional, epidemiological techniques into three syndromes; these are acute, chronic and subacute zinc deficiency. Acute zinc deficiency is relatively uncommon and follows parenteral hyperalimentation or oral L-histidine administration. Chronic zinc deficiency is more common, usually resulting from chronic dietary lack of zinc. Subacute or marginal zinc deficiency is the most common of these syndromes. Based upon a recent survey it is estimated that there are 4 million people in the U.S. with this syndrome, the initial symptom being dysfunction of taste and olfaction whereas treatment with exogenous zinc usually requires months before these symptoms return to normal. Diagnosis of these disorders was most efficacious following oral administration of Zn^{65} with subsequent evaluation of the kinetics of transfer of the isotope into various body tissues, the formulation of the data by compartmental analysis and the integration of the data by a systematic model of zinc metabolism.

Zinc deficiency is reflected in clinical syndromes which affect men and women of all ages and all socioeconomic and cultural classes in the U.S. It is neither prevalent in any specific area of the U.S. nor is it associated with any specific or definitive biochemical marker, which can make its identification difficult and at times frustrating and confusing. Its presence is manifested by a wide spectrum of symptoms, from acute, life threatening problems to mild subclinical or marginal disorders which may only vaguely disturb the well being of people who suffer with these complaints. The acute problems are often seen in profoundly ill patients treated in hospital whereas marginal or subclinical problems may be so vague that patients seek assistance outside traditional medical practice from practitioners employing hypnosis, chiropody or all encompassing nutritional therapies.

0097-6156/83/0210-0083$07.00/0
© 1983 American Chemical Society

These characteristics describe a protean set of disorders whose prevalence is unclear, whose symptomatology may vary from patient to patient and which do not exhibit specific biochemical markers or clinical set points by which to establish their relevance to specific pathological processes. They also point out some of the difficulties which clinicians face in dealing with zinc deficiency syndromes.

An example may help clarify these diagnostic difficulties. Iron deficiency has been recognized as a common problem in the U.S. for some time and its definition usually presents little difficulty since the microcytic, hypochromic anemia associated with its presence can be observed through the use of simple and commonly applied laboratory tests (i.e., evaluation of a red blood cell smear or measurement of blood hemoglobin and/or hematocrit). Although symptoms associated with iron deficiency may be vague or misleading (e.g., fatigue, nonspecific malaise, inability to concentrate for prolonged periods) the availability of simple and specific diagnostic laboratory tests make the evaluation of this problem straightforward and within current standards of practice of clinical medicine. Administration of exogenous iron is commonly associated with the correction of the abnormal laboratory tests resulting in remission of the associated symptoms.

Such is not the case with zinc deficiency. While symptoms of zinc deficiency can be as vague as those in iron deficiency, there is no simple diagnostic test which defines the condition. Zinc concentrations in blood, urine or hair commonly do not accurately reflect body zinc status (1-3). Although zinc is the major trace element found in fixed body tissues less than 1% of the approximate 2.5 g of total body zinc is circulating in blood (4) while for iron, the major trace element found in the circulation, over 50% of the approximate 4.0 g of total body iron is found in blood and blood forming systems (5).

Because zinc has multiple functions and is widely distributed in the body, its deficiency can affect many body tissues. In addition, effective treatment of zinc deficiency may require prolonged administration of large amounts of the metal with specific counter ions, a strategy which may not be apparent to physicians unfamiliar with these disorders.

While zinc deficiency syndromes have been known for many years, and their prevalance noted in diverse groups of people world wide, only recently have any identifiable patterns of some of these syndromes been formulated and the underlying mechanisms stated, however tentatively. These syndromes can be complex, present both in children and adults of both sexes and include single or multiple organ systems. These complexities can cause difficulty for both patients and physicians seeking an understanding of these processes.

Laboratory tests used to define these syndromes also may be confusing or inadequate (1-4). Urinary zinc excretion may be lower than normal, at normal levels, or elevated, as much as 3-10

fold normal. Serum, plasma and erythrocyte zinc concentrations
may be normal, low or high. Hair zinc levels may be lower than
normal or normal. Zinc treatment has not always provided a useful method for assessing deficiency, even retrospectively. Treatment embodying replacement of lost zinc has commonly produced a
remission of many of the symptoms of these syndromes provided body
zinc was restored to normal. However, oral administration of zinc
may not be of value in treatment of these disorders if other aspects of zinc metabolism are abnormal, e.g., zinc binding or
storage in tissues or transport in blood. Indeed, controlled
clinical trials of the efficacy of zinc treatment have not yet
been carried out for several specific disorders of zinc metabolism.

In an attempt to formulate our present knowledge about these
disorders it seemed useful to define some of the clinical and biochemical markers which accompany these syndromes even though they
may be neither specific nor diagnostic. To do this, we have attempted to re-evaluate zinc deficiency in its broadest sense and
to organize it into major diagnostic classifications using traditional, clinical techniques. In so doing our hope is to elucidate the incidence, symptomotology and characteristics of these
syndromes.

Clinical Definitions of Zinc Deficiency Syndromes

Acute Zinc Deficiency   Acute zinc deficiency is the most
easily identifiable of the zinc deficiency syndromes. This is a
relatively uncommon form of the syndrome occurring experimentally
following the oral administration of L-histidine (6) or clinically, following parenteral hyperalimentation (7-12) with the intravenous administration of high concentrations of glucose and free
amino acids (Table I). Symptoms of this disorder are not diagnostic, and it is easy to understand the dismay of a physician
viewing a young infant with an erythrematous, intracrural rash
unresponsive to antibiotic or antifungal therapy which disappears
within 24 hours after initiation of zinc treatment. The diagnosis
becomes somewhat simplier, if, in addition to the erythrematous,
intracrural lesions, the patient also exhibits paranychia, diarrhea, scattered bullous lesions and impaired healing of cuts and
scratches but these additional symptoms may not be present.

Acute zinc deficiency in adults is usually associated with
specific signs and symptoms associated with the rapid and progressive loss of body zinc. The initial symptoms of this condition is almost uniformly anorexia (6,13), a symptom also seen
early in experimental zinc deficiency in animals made deficient
by limitation of dietary zinc (14,15). Anorexia is commonly
followed by taste and smell dysfunction, then hypogeusia and hyposmia (loss of taste and smell acuity, respectively), followed
by dysosmia and dysgeusia.

Increased urinary zinc excretion has been observed from the
first day of this syndrome with as much as 1% of the total body

TABLE I

ACUTE ZINC DEFICIENCY

ETIOLOGY: Oral L-Histidine (16-64 gm)
                             Parenteral Hyperalimentation

ONSET: 3-14 days

INCIDENCE: Infrequent

SYMPTOM APPEARANCE   ZINC CHANGES

    Anorexia   ↑ Urinary Zn
    Hypogeusia
    Hyposmia
    Dysgeusia   ↓ Serum Zn
    Dysosmia   ↓ Zn-65 Retention
    Mental Confusion
    Cerebellar Dysfunction
    Ataxia
    Rash
    Buccal Lesions
    Acute Toxic Psychosis

zinc excreted daily (6). Decreased serum zinc concentration
usually occurs later but it is small in relation to the marked
elevation of urinary zinc excretion. Oral administration of Zn^{65}
to patients with acute zinc depletion has been associated with decreased retention (13) of the isotope indicating a measureable
loss of body zinc rather than simply body zinc redistribution.

Persistent acute body zinc loss has resulted in many symptoms
in addition to taste and smell dysfunction if losses continue at
significant levels for periods of 14 days or more. These symptoms
include mental confusion, cerebellar dysfunction, including inten-
tion tremor and ataxia, an erythrematous, intracrural rash, buc-
cal epithelial lesions and ulcers, and acute toxic psychosis (6).
The altered mental state of these patients can be related to the
rapid depletion of the relatively high zinc content in the limbic
system of the brain (16,17), demonstrating that zinc can cross
the blood brain barrier bidirectionally, dependent upon the
gradient, and that it can be readily mobilized from brain tissue.

Treatment of this disorder with oral or parenteral zinc is
almost uniformly associated with the correction of all aspects of
this disorder. Even with continued administration of L-histidine
large amounts of exogenous zinc were effective in correcting the
signs and symptoms of this disorder (6) with the anorexia, taste
and smell dysfunction, ataxia and mental symptoms returning to
or toward normal within 24 hours, the skin or epithelial lesions
requiring days of therapy (6). Cessation of L-histidine administration for 2-4 days was not successful either in producing symptom remission (6) or in returning body zinc levels to normal.
It was only after administration of exogenous zinc that the signs
and symptoms of this disorder were corrected, constituting replacement of lost body zinc stores.

In infants, commonly neonates given parenteral hyperalimentation, the erythrematous, intracrural rash, paranychia and occa-
sional bullous skin lesions are early signs of this disorder, although the inability to console the infant may be the earliest,
most sensitive, yet most difficult to evaluate, symptom of this
disorder {Table 1, (18)}. Bacterial and fungal cultures of the
skin lesions are often negative and there is usually no associated febrile condition (18). Serum electrolytes and blood gases
are commonly within normal limits. Serum zinc levels may be
lower than 10 µg/dl and may provide the simplest and most useful
laboratory test of this disorder (18-20). Oral or parenteral
zinc administration is usually associated with a rapid remission
of these symptoms although changes in serum zinc concentration
may lag the clinical changes observed (18).

Increased urinary zinc excretion coupled with decreased serum
zinc concentration and the observed symptoms follow acute zinc
depletion with enough consistency that it is relatively simple for
an alert physician to establish the diagnosis. However, it is
this awareness and the protean nature of the symptoms which complicate its diagnosis. Indeed, the full-blown clinical picture

of zinc deficiency either in infants or adults can, as noted, be
confusing and mistaken for several other clinical disorders.

   Chronic Zinc Deficiency   Chronic zinc deficiency is more
common than acute zinc deficiency occurring in adults experimentally following administration of zinc deficient diet or naturally
following the intake of a diet low in zinc (21-23). It also has
been observed in children who commonly do not eat zinc rich foods
(24,25). Children with this condition may exhibit small stature
(oftimes below the 5th percentile for height and weight) and poor
school performance {(25) Table II}. Infants with this condition
can exhibit genetic{(26,27) acrodermatitis enteropathica)} or ac-
quired (28-30) forms of this disorder, which occur in some patients following substitution of cow's milk for human milk in the
neonatal or perinatal period (29,30). Chronic zinc deficiency
was first reported in Iran and Egypt (22,23) in adults who exhibited short stature, hepatosplenomegaly, iron defiency anemia
and hypogonadism. These symptoms were, in part, related to the
intake of zinc ligands which appeared to bind ingested zinc pref-
erentially, making the metal less available for absorption in the
gastrointestinal tract (31). Dwarfism was also observed in these
patients (22,23). Phytate (31), oxalates (32), and forms of fiber (33,34) have been suggested as ligands that could bind zinc
leaving it relatively unavailable for effective gastrointestinal
absorption. Other factors, including pica, and intestinal para-
sites, may have been responsible for some aspects of the anemia
and hepatosplenomegaly reported in the Iranian and Egyptian patients with this disorder (22,23). Some investigators have considered the presence of putatively more effectively absorbable
zinc ligands in human milk than that in cow's milk responsible
for the delayed appearance of acrodermatitis enteropathica in
infants, the symptom complex becoming apparent only after the
introduction of cow's milk into the diet (29,30,36-39). These
latter observations have spawned considerable controversy between
proponents of citrate (36,37) or picolinate (38,39) as the major
zinc ligands in human milk.
   Children may exhibit different symptoms than adults at the
onset of this disorder. In children, anorexia, diarrhea, ir-
ritability and short stature may be apparent whereas in adults
taste and smell dysfunction, hypogonadism and poor wound healing may appear as early signs. Although taste and smell dysfunction are common components of this syndrome they may be over-
shadowed by the more acute symptoms of diarrhea and lethargy.
   A number of important and, as yet, unexplained discrepancies
are apparent in tissue zinc concentrations in this syndrome. Patients with chronic renal disease, with clinical signs and symptoms of zinc deficiency show consistently elevated red blood
cell zinc levels (40-42) whereas plasma zinc concentrations have
been reported as either low, normal or elevated (40-44). Patients with Kwashiorkor and symptoms of zinc deficiency have

TABLE II

CHRONIC ZINC DEFICIENCY

ETIOLOGY: Dietary Lack
                 Absorption Inhibition
                 Abnormal Zn Ligand
                 Chronic Renal Failure
                 Cancer

ONSET: Months, Years

INCIDENCE: Moderate (10% of growth retarded children,
                           5th centile)

SYMPTOM APPEARANCE	ZINC CHANGES
Children	
Short Stature	↓ Hair Zn
Hypogeusia	↓ Serum Zn
Pica	↓ Salivary Zn
Hyposmia	↓ Urinary Zn
Paranychia	↓ RBC Zn
Rash	↓ Zn-65 Absorption
Diarrhea	
Adults	
Hypogeusia	↓ Hair Zn
Hyposmia	↓ Serum Zn
Hypogonadism	↓ Urinary Zn
Short Stature	↓ RBC Zn
Diarrhea	↓ Zn Enzymes
Poor Wound Healing	↓ Serum T, LH

serum zinc levels that are lower than normal (45,46) but hair zinc
levels which are normal or even elevated rather than below normal
as might be expected initially (47). Urinary zinc excretion may
be lower than normal if dietary zinc is low, but higher than normal if the anorexia associated with this syndrome is prominent, related to the effects of chronic weight loss (48). In several disorders of chronic zinc deficiency, serum or plasma zinc levels
may be normal, high or low and thus of limited diagnostic value
(1-3,49). Some zinc enzymes, plasma testosterone or LH may be
decreased in patients with hypogonadism or with azospermia (50).
Cellular immunity may also be impaired in patients with chronic
uremia (51). Studies of oral absorption of Zn^{65} in patients
with acrodermatitis enteropathica have produced results consistent with impaired absorption (52) with total body zinc levels
suspected to be lower than normal (53). However, Zn^{65} absorption
in these patients overlaps the lower end of the normal absorption
range, a phenomenon which has only recently been emphasized (54,
55).
    The clinical diagnosis of chronic zinc deficiency is simpli-
fied if zinc is depleted from most body tissues. However, the
long time period necessary for the appearance of symptoms of
this syndrome and its protean nature can complicate its clinical
diagnosis, especially if physicians do not consider its presence.
    Treatment with exogenous zinc has generally resulted in suc-
cessful remission of the symptoms of this syndrome. However, a
remarkably small amount of zinc is necessary to restore normal
function in patients with acrodermatitis enteropathica, often only
slightly more than the recommended daily dietary allowance for
zinc (56). Diodoquin, a drug which appears to assist zinc absorption, has also been useful in the treatment of some patients
with acrodermatitis enteropathica (57). In some patients, larger
amounts of zinc given for prolonged periods of time are necessary
to produce symptom remission. In the only double blind study performed in patients with putative zinc deficiency in Egypt, placebo
was as effective as zinc in treating the disorder (58). These
results offer a confusing picture of response to therapy which
emphasizes some of the difficulties with the clinical appreciation
of these problems.
    Subacute Zinc Defiency Subacute zinc deficiency is the ma-
jor clinical zinc deficiency syndrome in the U.S. Based upon a
national survey of smell function performed in 1980, estimates
suggest that this syndrome may affect as many as 4 million people
(59). The etiology of subacute zinc deficiency is related to a
variety of disorders which do not, at first glance, appear to
have any specific relationship to a metabolic disease process
{(Table III) (60)}. Indeed, the mechanisms by which a viral ill-
ness, head injury, surgical procedures, or allergic rhinitis may
produce lowered body zinc levels are still unknown. In addition,
to complicate this problem further, the onset of this syndrome
can occur relatively quickly, over days, or more slowly, over

TABLE III

SUBACUTE ZINC DEFICIENCY

ETIOLOGY: Dietary Lack
                 Viral URI (PIHH)
                 Head Injury
                 Unknown

ONSET: Days, Weeks, Months

INCIDENCE: 4 million in U.S.

SYMPTOM APPEARANCE	ZINC CHANGES
Hypogeusia	↓ Serum Zn
Hyposmia	↓ RBC Zn
Dysgeusia	↓ WBC ALK PO_4'ase
Dysosmia	↓ Zn-65 Absorption
	↓ Total Body Zn

weeks or months. The symptoms of this disorder are as protean
as those of the acute or chronic zinc deficiency syndromes and
provide no special diagnostic clues to the etiology of these
problems. However, the most obvious symptoms associated with
this most common form of zinc deficiency are taste and smell dysfunction, hypogeusia, hyposmia, dysgeusia and dysosmia (60).
These are the major, and oftimes the only, symptoms associated
with this disorder. Although acute and chronic zinc deficiency
syndromes can also be characterized by taste and smell dysfunction they also have other, more easily recognized dysfunctions
affecting skin, gastrointestinal tract, neuromuscular system or
other body systems allowing their diagnosis to be made somewhat
easier. Of all the zinc deficiency syndromes only subacute zinc
deficiency can present with taste and smell dysfunction as the
major or only set of clinical symptoms manifested by the condition.

    Thus, subacute zinc deficiency may represent one of the
most difficult diagnostic challenges of all zinc deficiency syndromes. Although serum and erythrocyte zinc concentrations may be
significantly lower than normal in these patients (61), these
levels may also be well within normal limits. Urinary zinc excretion may be normal, elevated or depressed (62). Salivary zinc
concentration is usually below the normal mean but it may also be
within the normal range (64,65). Measurements of salivary gustin
can be useful in some patients (3,49) but it is not necessarily
diagnostic (3,49). Measurements of zinc absorption are also not
diagnostic for after oral administration of Zn^{65} absorption has
varied over a wide range, from 17 - 100% of the administered dose
(3,49,60). These results emphasize the variability of zinc concentration in body tissues in this syndrome and the lack of a single,
specific functional test by which these patients may be classified
as zinc deficient. Indeed, because these results are so variable
the classification of these patients as zinc deficient may, at
first glance, appear to be questionable.

    This variability in single tests is not apparent, however,
when specific tests of tissue and body zinc status are examined
simultaneously. Thus, measurements of impaired zinc absorption
(3,49,60) and lower than normal concentrations of tissue and body
zinc in patients with this disorder (66-69) confirm the diagnosis
of subacute zinc deficiency with certainty (Tables IV, V). The
rationale for this classification was derived from studies of
compartmental analyses of the kinetic data obtained from these patients following oral Zn^{65} administration, measurement of zinc
retention in several body tissues, including areas over the liver
and thigh, in red blood cells, plasma, urine and stool, and by
synthesizing these data by means of a mathematical model in which
all of these data could be constrained and fitted simultaneously
(67,69). By carrying out these processes we were able to calculate parameters which were both necessary and sufficient to identify these patients. We were also able to measure lower than

normal tissue zinc concentrations in some of the patients (Table V). In order to understand the basis of this formulation specific questions about the nature of zinc deficiency had to be raised and answers attempted. Only through these formulations could awareness of the issues underlying zinc deficiency be clarified.

Critical Questions of Zinc Deficiency

A series of questions raised by the above ambiguities may be helpful to formulate a redefinition of zinc deficiency. Initially four questions can be raised. They are: (1) How can zinc deficiency be defined? (2) Why do patients with subacute zinc deficiency, exhibit taste and smell dysfunction as their major, often their only, symptom complex? (3) What is the meaning of the wide range of zinc absorption in patients with subacute zinc deficiency?, and (4) Does zinc treatment of patients with subacute zinc deficiency adequately correct the clinical symptoms of the disease processes, and if successful, by what mechanisms do these changes occur?

1. Definitions of Zinc Deficiency. It is clear that zinc deficiency can be defined only with difficulty since the zinc content of a single body fluid or tissue level cannot provide a definitive estimate of body zinc status (1-3). It is also not possible to rely on the concentration of zinc in several body tissues to estimate body zinc status since differing concentrations in different tissues lead to non-definitive conclusions.

With these negatives in view it seemed important to determine if zinc deficiency could be defined quantitatively by means of any set of laboratory tests. In an attempt to clarify this, a series of studies were carried out following oral administration of Zn^{65}, after an overnight fast, in normal volunteers (55) and in patients with subacute zinc deficiency (3,49,60,66). As noted above, mean zinc absorption in patients with taste and smell dysfunction was 55% ranging from 17% to 100% (3,49,60), whereas in normals absorption varied from 42 to 84% with a mean of 65%, a value significantly higher than that measured in the patients (55,60). If zinc deficiency were similar to iron deficiency, then a lowered total body zinc content should be expected to yield results similar to those observed with low total body iron content; i.e., an increased absorption of zinc associated with decreased body zinc stores. However, this is not the case for human zinc deficiency since mean absorption of zinc in these patients is significantly lower than normal (3,49,60) in the face of what appears to be a lower than normal total body zinc status (60,69). These results, in patients with taste and smell dysfunction, are consistent with results obtained in similar studies in patients with cancer (55), dermatological (55), and other disorders (56) and raise fundamental problems about mechanisms of human zinc deficiency. These discrepancies in human zinc metabolism are not apparent in the rat in which decreased body zinc stores have been associated with

TABLE IV

COMPARISON OF ABSORPTION AND EXCRETION OF Zn-65
IN NORMAL VOLUNTEERS AND IN PATIENTS WITH SUBACUTE Zn DEFICIENCY

CONDITIONS	NORMALS	PATIENTS
$V_i{}^+$	.146 ± .005	.150 ± .007
V_u	.009 ± .001	.007 ± .001
V_F	.138 ± .006	.144 ± .007
V_a	.36 ± .05	.11 ± .02*
V_r	.35 ± .05	.10 ± .01*
∞**	.69 ± .03	.43 ± .04
$L_u{}^{++}$	.26 ± .04	.18 ± .03
L_F	2.7 ± .2	1.4 ± .2*
$L_{(15,24)}$	73 ±13	26 ±3*

+   mg/kg/day

++ day^{-I}

* $p < 0.01$

**   fraction of gastrointestinal Zn absorbed

V   is defined as the mass transport in units of mg/day

V_i rate of zinc transfer from C_{15} to C_{24} calculated as the product of $L(24,15)$ (M_{15})

M   is defined as the mass, in mg, of compartment C_i

L_{ij} is defined as the fraction of zinc in compartment C_j transferred to compartment C_j per unit time. Its units are day^{-1} and represent the rate constant from compartment j into compartment i.

V_u, V_F total zinc excretion rate into urine and feces, respectively. $V_F = L(0,25)(M_{25})$.

V_a rate of zinc absorption and is $L(15,24)(M_{24})$

V_r rate of zinc reabsorption

∞   the fraction of zinc absorbed in the gut.

$$= \frac{V_a}{V_i + V_r} = \frac{V_{max}^{GI}}{K_m^{Gi} + (V_i + V_r)}$$

Where K_{max}^{GI} is the maximum rate of zinc absorption

and K_m^{GI} is an effective Michaelis Menten equilibrium constant

L_u = Vu/M_{15}

L_F = $(1 - ∞)(L24,15)$

increased zinc absorption and increased body zinc stores associated with decreased zinc absorption (cf, Weigard, E. and Kirchgessner, M., Homeostatic adjustments in zinc digestion to widely
varying zinc intake, Nutr. Metab. 22:101-112, 1978). In addition,
the decreased body zinc stores in the rat have been associated with
decreased activity of several zinc dependent enzymes, a condition
which is not observed uniformly in humans.

One simplistic manner of dealing with these data is to assume
some gastrointestinal block to the absorption of zinc which would
account for the symptoms of the patients as well as the lower than
normal absorption. This would be an acceptable solution provided
all other data were consistent, which they are not. Serum zinc
concentrations in these patients extended over a wide range as did
urinary zinc excretion. Both of these values should be below normal if zinc absorption were the major or only defect in these patients. Similarly, red blood cell zinc concentrations extend over
a wide range, albeit mean levels are lower than normal (61). This
latter finding becomes more difficult to explain when it is
learned that treatment with large amounts of exogenous zinc does
not increase these red blood zinc levels although serum zinc concentration and urinary zinc excretion both rise significantly (61).

In order to deal with these complex problems all data from
the oral Zn^{65} studies obtained in patients with taste and smell
dysfunction were organized and submitted to compartmental analysis (68,69) with the subsequent development of a model (Figure 1)
which accounted for all the data obtained over the entire period
of these studies, both prior to and after treatment with exogenous
zinc (69). These results, compared in normal volunteers, demonstrated that not only was absorption of zinc significantly impaired in the patients compared with the normal volunteers (Table
IV) but also that the rate at which zinc was absorbed was significantly lower in the patients than in the normals (3,55,60) and
that their total body level of zinc was lower than in the normals
(60,69). By the use of this model it was also possible to specify
those conditions which were both necessary and sufficient to
identify patients with zinc deficiency (60,69). With these techniques it was possible to identify, by objective criteria, laboratory tests by which patients with subacute zinc deficiency could
be defined quantitatively. It was also possible to measure various
tissue and total body zinc levels and to compare patients with
normals so that patients with zinc deficiency could be identified.
The major problems presented with these techniques are that they
are time consuming, cumbersome, expensive and are presently unavailable in many areas of the U.S.

2. Why are taste and smell dysfunction the major clinical
problems in patients with subacute zinc deficiency? It is important to recognize that subacute zinc deficiency is usually a mild
or marginal form of the deficiency. It is also important to recognize that the earliest signs of both acute and chronic zinc deficiency, i.e., during the period when the deficiency is beginning
to be manifested, are those of taste and smell dysfunction. These

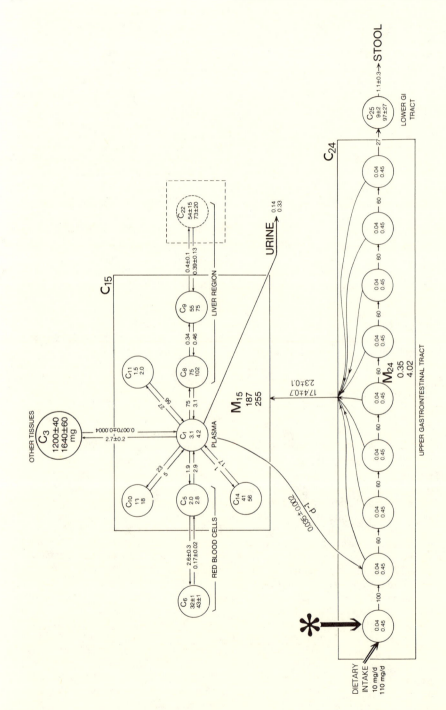

Figure 1. In this model of zinc metabolism in humans circles represent compartments and arrows represent transfer pathways. (Reproduced with permission from Ref. 69.)

The circles are labeled C_i for the C_i^{th} compartment. The arrows represent rate constants, the $L_{(1)}$ defined as the fraction of zinc in compartment C_1 transferred to compartment C_1 per unit time. The plasma, red blood cell, and liver subsystems are indicated by C_1, C_5, C_6, and C_8, C_9, and C_{28}, respectively. C_{22} in the liver subsystem, which is set off by dotted lines, did not appear in the original, short term model (67) but was required in the extended model as explained in the text. Thus, the circles and arrows depicted here, except for those related to C_{22}, are those defined in the original short term model (67). The rectangles illustrate the result of reducing the original, short term model. The upper rectangle includes all of the rapidly turning over compartments defined in the original short term model. This rectangle shows the constituents of C_{15} in the reduced model. The lower rectangle includes the entire upper gastrointestinal tract and is called C_{41} in the reduced model. C_5, C_6, C_{28}, and C_{35} are slowly turning over compartments and they are kept in this form in the reduced model. The numbers shown in this figure are from a representative patient, patient 4, in the group reported. The upper numbers indicate data obtained before, the lower number during, oral zinc loading. If only one number appears, no change occurred during zinc loading. To exemplify the rate constant, the arrow from C_1 to C_8 represents $L_{(8,1)}$, which has a value of 75/day. The rate constants are reported with and without SD. Those without SD are taken from the original short term model where SD were shown; these numbers are included here simply to show the basis for combining the compartments into a single composite compartment as explained in the text. Those rate constants with SD are derived from the data obtained in this representative patient and were fitted to the reduced model. To exemplify the masses, the mass of the plasma compartment C_1 is 3.1 mg before, and 4.2 mg during, oral zinc loading. Numbers without SD are the masses in the rapidly turning over compartments; the sum of these numbers is the mass of the reduced compartment described above. Masses with SD were derived from the data obtained in this representative patient and were fitted to the reduced model. ^{65}Zn is introduced into the first compartment of the upper gastrointestinal tract as indicated by the asterisk. Dietary intake is indicated by the double arrow.

results suggest that those zinc pools subserving the taste and
smell systems are small, labile, and in rapid equilibrium with
zinc intake. Thus, with too little intake or with early depletion of zinc the first systems to be affected are those subserving taste and olfaction.

    3. Why is there such a wide range of zinc absorption in
patients with subacute zinc deficiency? While the answer to this
question is unclear, the data suggest that zinc absorption and its
distribution in the human body is unique and differs significantly
from iron absorption or that of other divalent metal ions. Some
data relevant to this question were obtained from the zinc loading
studies carried out in our laboratory in patients with subacute
zinc deficiency (68,69). If their oral zinc load increased 10 fold
the zinc fraction absorbed decreased by a factor of 5 (from 43% to
9%) with absorption increasing by a fraction of two (from 4.5 g/d
to 10.6 g/d). Presently, data from normal volunteers are being
analyzed under the same conditions so that a direct comparison can
be made. Data obtained by the use of the model allows further insight into this problem. Ninety percent of total body zinc in normal volunteers and in patients with taste and smell dysfunction are
found in one large compartment (C_3) which represents, primarily,
zinc in muscle and partly zinc in bone (60,69). It is the largest
body pool of zinc. The rate constants into and out of this compartment from the rapidly exchanging body zinc pool (C_{15}) which
includes plasma are both very slow. These results suggest that the
zinc content in this compartment and the rate constants into and
out of this compartment are the controlling factors not only in zinc
absorption but also in other aspects of zinc metabolism in humans.
Thus, some of the confusion about zinc absorption may be related,
in part, to differences in steady state zinc content as well as
the rates of zinc accumulation and loss in this pool.

    These concepts emphasize the need to consider several factors
prior to establishing a diagnosis of zinc deficiency. Not only
is it necessary to know zinc absorption through the gastrointestinal tract but also the type of zinc binding ligands in the gastrointestinal tract needed to transport zinc across the gut mucosal surface, the presence of appropriate zinc binding proteins
and the presence of tissue specific zinc storage proteins to note
just a few. Differences between gastrointestinal absorption of
iron and zinc in humans and local gut factors which may influence the
absorption of these metals were recently discussed (Matseke, J.W.,
Phillips, S. F., Malagelada, J. R. and McCall, J. T. Recovery of
dietary iron and zinc from the proximal intestine of healthy man:
Studies of different meals and supplements. Amer. J. Clin. Nutr.
33:1946-1953, 1980) and emphasize the importance of dietary factors which can influence human zinc absorption. Only with an understanding of these and other factors can zinc deficiency be defined
clearly. Presently, such an awareness is far from complete.

    4. Does treatment with zinc correct the taste and smell dys-
function observed in patients with subacute zinc deficiency? This

question can be answered only providing that specific requirements
and criteria are met. First, prolonged daily treatment for 2–4
months with high doses of zinc (100 mg of zinc ion) is usually required prior to reversal of these symptoms. An explanation for
this prolonged time period may relate to the hypothesis that compartment C_3 must be saturated with zinc before the symptoms remit
in spite of the lability and small size of the zinc pools subserving the taste and smell systems. Thus, as in an old clinical
saw, those systems which lose function first may also be those which
regain function last when appropriate treatment is given. These
hypotheses are based, as noted previously, upon the slow rate of
uptake and release of zinc from this major body zinc pool. Second,
these results must take into consideration the negative results of
the double blind study we published in 1976 (63). In this study
placebo was as effective as zinc in restoring taste and smell to
normal patients with taste and smell dysfunction. The important
consideration here is that recent studies indicate that no more
than 25% of patients with taste and smell dysfunction would be
expected to exhibit zinc deficiency (3,60) and that all placebo
responders analyzed in a subsequent study (3,60) were found in
the group of patients who were zinc sufficient. Thus, the prior
double blind study exhibited a serious flaw in that our knowledge
of body zinc status was severely limited both before and during
the study. This lack of knowledge, forced the inclusion into the
study of many patients who were not zinc deficient and who would
not be expected to be zinc responsive. The reasons for their re-
sponsiveness to placebo is still unclear (3,60,63) although several hypotheses have been advanced to explain this result (3,60).
Third, a subsequent single blind study was carried out in patients with proven subacute zinc deficiency (3,60) and positive
results with zinc treatment in correcting the taste and smell
dysfunction were obtained. In this study if only patients who
were relatively low absorbers of Zn^{65} (i.e., zinc absorption
more than 2SD below the mean level of absorption of normal volunteers) were evaluated then initial treatment with placebo was
shown to be effective in correcting their symptoms; subsequent
treatment with placebo produced a return to or toward functional
loss of taste and smell acuity. As a clinical technique to obtain some useful estimate of body zinc status saliva zinc concentration appeared helpful and, when this index was used as a measurement of treatment success or failure there was a significant
positive correlation both in the previous souble blind study (63)
as well as in the present single blind study (3,60). These results indicate that until there was correction of saliva zinc
concentration to normal there was no associated correction of the
hypogeusia (3,60,63). If no changes in saliva zinc occurred there
was no measurable improvement in taste acuity. These studies were
further supported by studies of gustin induction in patients with
hypogeusia and subacute zinc deficiency (60). Treatment with zinc
in these patients not only returned taste function to normal but

also corrected the low level of salivary gustin to normal (70).
Although these analyses provide useful data and important hypotheses definitive acceptance of zinc treatment for these symptoms
cannot be obtained until a double blind clinical trial is carried
out using these techniques and hypotheses. We have attempted to
mount such a study but have faced repeated difficulties in its
performance.

From the zinc model it was possible to compare the tissue
concentrations of zinc in the patients with normals. Results indicated that patients exhibited significantly lower levels of zinc
than normals in erythrocytes, liver and muscle, whereas zinc con-
concentration in their gastrointestinal tract was elevated above
normal (Table V). These results emphasize again the necessity to
obtain measurements of body zinc status prior to the definition
of zinc deficiency.

From the zinc model it was also possible to determine those
factors which were both necessary and sufficient to define subacute zinc deficiency. These included the rate constant of zinc
absorption, the rate constant of zinc exchange from C_{15} and C_3 in
the thigh (3,60). From these data it was possible to relate the
differences between the patients and normals to two major factors,
differences in (decreased) gastrointestinal Zn^{65} absorption and
decreased transfer of Zn^{65} into C_3 (Table VI). These results also
suggest two possible mechanisms for the zinc impairment observed
in subacute zinc deficiency. These are:

1. Down regulation of zinc carrier molecules in which
    patients adapt to less Zn in critical body pools;
    and/or

2. An increase in the effective Michaelis-Menten constant.

With our present knowledge it is not possible to choose between these two hypotheses.

Summary and Conclusions

These studies offer a reformulation of our present knowledge
of zinc deficiency in a traditional, clinical manner. This reformulation is similar to that used by most other investigators dealing with other clinical problems. This reformulation or redefinition on the basis of acute, chronic and subacute zinc deficiency,
has allowed for the classification of zinc deficiency syndromes
with respect to etiology, incidence, onset, symptomatology and
diagnostic procedures. This reformulation has also allowed for
the development of hypotheses by which to test the usefulness of
the definitions and to describe the characteristics which define
each of the subcategories.

It is clear that subacute zinc deficiency represents the most
common form of zinc deficiency encountered in the U.S. and that
both its diagnosis and effective treatment are the most difficult
of all the deficiency syndromes. The estimate of four million
people in the U.S. with this disorder suggests that it occurs
much more frequently than previously considered.

TABLE V

COMPARISON OF Zn CONCENTRATIONS IN VARIOUS TISSUES IN NORMAL
VOLUNTEERS AND IN PATIENTS WITH SUBACUTE Zn DEFIENCY

TISSUE CONCENTRATION (mg)	NORMALS	PATIENTS
Plasma	.032 ± .001	.039 ± .002
GI Tract		
M_{24}	.0051 ± .002	.0062 ± .003*
M_{25}	.35 ± .10	.32 ± .09
Erythrocytes		
M_6	.96 ± .06	.54 ± .05*
RBC Mass	.98 ± .06	.56 ± .05*
Liver Region		
M_{22}	2.0 ± .3	1.3 ± .2*
Liver Mass	5.3 ± .3	3.5 ± .5*
Thigh Region		
M_3	32 ± 3	15 ± 2*
Thigh Mass	3.8 ± .5	2.7 ± .3

M is defined as the mass in mg of zinc in compartment C_i

TABLE VI

TWO FACTORS ARE NECESSARY AND SUFFICIENT TO
DISTINGUISH NORMAL VOLUNTEERS FROM PATIENTS
WITH SUBACUTE ZINC DEFICIENCY

---

1. GASTROINTESTINAL Zn-65 ABSORPTION

$$\propto = \frac{V_a}{V_i + V_r} = \frac{V_{max}^{GI}}{K_{max}^{GI} + (V_i + V_r)}$$

V_a,   rate of Zn absorption

V_i,   rate of Zn ingestion

V_r,   rate of Zn recycling secretion from plasma into
             intestine

V_{max}^{GI},   maximum rate of Zn absorption

V_{max}^{GI},   effective Michaelis-Menten equilibrium constant

2. TRANSFER OF Zn-65 INTO C_3

     L (3,15)
     Fraction of C_3 in thigh

     Data are presented dealing with a zinc model by which total
body zinc mass and other parameters of zinc metabolism can be
measured quantitatively such that patients with subacute zinc deficiency can be differentiated from normals by a series of laboratory tests, albeit cumbersome and expensive. The major symptom
complex of subacute zinc deficiency is related to taste and smell
dysfunction and these symptoms are discussed relative to the body
distribution of zinc and the dysfunctions of zinc deficiency. Possible mechanisms by which zinc corrects the deficiency are discussed. These results present a new approach to the diagnosis and
treatment of zinc deficiency and help to explain the manner by
which these syndromes occur, can be diagnosed and treated.

Literature Cited

1. Sandstead, H.H. Amer. J. Clin. Nutr. 1973, 26, 1251-60.
2. Henkin, R. I. New Eng. J. Med. 1974, 291, 675-6.
3. Henkin, R. I.; Aamodt, R. L.; Foster, D.M. "The Clinical,

Biochemical and Nutritional Aspects of Trace Elements";
Alan R. Liss, Inc., New York, NY, 1982 (In Press)

4. Henkin, R. I.; Apgar, J.; Cole, J. F.; Coleman, J. E.; Cotterill, C. H.; Fleisher, M.; Goyer, R. A.; Greifer, B.; Knezek, B. D.; Mushak, P.; Piscator, M.; Stillings, B. R.; Taylor, J. K.; Wolfe, D. A. "Zinc"; University Park Press, Baltimore, MD, 1979; p 123-72.

5. Finch, C. "Iron"; University Park Press, Baltimore, MD, 1979; p 79-106.

6. Henkin, R. I.; Patten, B. M.; Re, P.; Bronzert, D.A. ARch. Neurol. 1975, 32, 745-51.

7. Van Rij, A.; McKenzie, J.M. "Trace Elements in Man and Animals III"; Arbeitskresi fur Tierernahrungsforschung, Weihenstephan, Freising-Weihenstephan, West Germany, 1978; p 288-91.

8. Lowry, S. F.; Goodgame, Jr., J.T.; Smith, Jr., J.C.; Maher, M.M.; Makuch, R.W.; Henkin, R.I.; Brennan, M. F. Ann. Surg. 1979, 189, 120-8.

9. Arakawa, T.; Tamura, T.; Igaraski, Y. Amer. J. Clin. Nutr. 1976, 29, 197.

10. Fleming, C. R.; Hodges, R.E.; Hurley, L. S. Amer. J. Clin. Nutr. 1976, 29, 70.

11. Hankins, D. A.; Riella, M. D.; Scribner, B. H.; Bragg, A. L. Surg. 1976, 79, 674.

12. Kay, R. G.; Tosman-Jones, C.; Pylos, J. Ann. Surg. 1976, 183, 331.

13. Henkin, R. I. Ann. N.Y. Acad. Sci. 1977, 300, 321-34.

14. Chesters, J.K.; Quarterman, J. Brit. J. Nutr. 1973, 30, 553-66.

15. McConnell, S. D.; Henkin, R.I. J. Nutr. 1974, 104, 1108-14.

16. Donaldson, J.; St. Pierre, T.; Minnich, J.L.; Burberi, A. Can. J. Biochem. 1973, 51, 87-92.

17. Crawford, I. L.; Connor, J. D. J. Neurochem. 1972, 19, 1451-8.

18. Sivasubramanian, K. N.; Henkin, R. I. J. Ped. 1978, 93, 847-51.

19. Hambidge, K.M. "Trace Element Metabolism in Animals II"; University Park Press, Baltimore, MD, 1973, p 171-183.

20. Snita, S.; Ikedu, K.; Magusak, A.; Hayosheda, Y. J. Pediatr. Surg. 1978, 13, 5.

21. Prasad, A. S. "The Clinical, Biochemical and Nutritional Aspects of Trace Elements"; Alan R. Liss, Inc., New York, NY, 1982 (In Press)

22. Prasad, A. S.; Mulsted, J. A.; Nadimi, M. Amer. J. Med. 1961, 31, 532.

23. Prasad, A. S.; Miale, Jr.; A.; Farid, Z.; Sandstead, H.H.; Schulert, A.R. J. Lab. Clin. Med. 1963, 61, 537.

24. Hambidge, K.M.; Hambidge, C.; Jacobs, M.; Baum, J.D. Pediatr. Res. 1972, 6, 868-74.

25. Hambidge, K.M.; Walravens, P. A. "Trace Elements in Human Health and Disease, I"; Academic Press, New York, NY, 1976, p 21-32.

26. Danbolt, N.; Closs, K. Arch. Derm. Venereal. 1942, 23, 127-69.

27. Moynahan, E. J. Lancet 1974, 21, 399-400.

28. Moynahan, E. J. and Barnes, P. W. Lancet 1974, 21, 399-400.
29. Entwisle, B.G. Austral. J. Derm. 1965, 8, 13-21.
30. Sunderman, Jr., F. W. Ann. Clin. Lab. Sci. 1975, 5, 132-45.
31. Oberleas, D.; Muhrer, M. E.; O'Dell, B. L. "Zinc Metabolism";
    C. C. Thomas, Springfield, IL, 1966,
32. Kelsey, J. This volume
33. Reinhold, J. G., Parsa, A., Karimian, N., Hammick, J. W.;
    J. Nutr. 1974, 104, 976-982.
34. Aamodt, R. L.; Rumble, W.F.; Johnston, G. S.; Markley, E.S.;
    Volpe, T.; Henkin, R. I. Fed. Proc. 1980, 39, 651.
35. Hambidge, K. M.; Silverman, A. Arch. Dis. Child. 1973, 48,
    567-568.
36. Hurley, L. S.; Lunnerdal, D. B. Nutr. Rev. 1980, 38, 295.
37. Hurley, L. S.; Lunnerdal, D. B. Ped. Res. 1981, 15, 166-7.
38. Evans, G. W. Nutr. Rev. 1980, 38, 137-41.
39. Evans, G. W.; Johnson, E. C. J. Nutr. 1980, 110, 1076-80.
40. Mansouri, K. J.; Halsted, A.; Gumbos, E. A. Arch. Intern. Med.
    1970, 125, 88-93.
41. Lindeman, R. D.; Baxter, D. J.; Yunice, A. A.; Kims, R. W.,
    Jr.; Kraikit, S. Prog. Clin. Biol. Res. 1977, 14, 193-209.
42. Sandstead, H. H. Amer. J. Clin. Nutr. 1980, 33, 1509-1516.
43. Mahajan, S. K.; Prasad, A. S.; Rabbani, P.; Briggs, W. A.;
    McDonald, F. D. J. Lab. Clin. Med. 1979, 693-698.
44. Bogden, J. D.; Oleske, J. M.; Weiner, B.; Smith, L. C., Jr.;
    Smith, L. C.; Majem, G. R. Amer. J. Clin. Nutr. 1980, 33,
    1088-95.
45. Hansen, J. D. L.; Lehmann. S. Afr. Med. J. 1969, 43, 1248-51.
46. Sandstead, H. H.; Shukry, A. S.; Prasad, A. S.; Gabr, M. K.;
    El Hifney, A.; Mokhtar, N.; Darby, W. J. Amer. J. Clin. Nutr.
    1965, 17, 15-26.
47. Kopito, I.; Schwachman, H. "The First Human Hair Symposium";
    Medcom Press, New York, NY, 1974, p 83-90.
48. Spencer, H.; Samachson, J."Trace Element Metabolism in
    Animals"; Edinburgh, Scotland, 1970, p 312-14.
49. Henkin, R. I.; Aamodt, R. L.; Babcock, A. K.; Agarwal, R. P.;
    Shatzman, A. R. "Perception of Behavioral Chemicals"; El-
    sevier/North Holland Biomedical Press, Amsterdam, The Nether-
    lands, 1981, p 227-65.
50. Mahajan, S. K.; Prasad, A. S.; Briggs, W.A.; McDonald, F. D.
    Trans. Amer. Soc. Artif. Intern. Org. 1980, 26, 133-138.
51. Antonion, L. D.; Shalbomb, R. J.; Schechter, G. P. Amer. J.
    Clin. Nutr. 1981, 34, 1912-17.
52. Lombeck, I.; Schnippering, H. G.; Ritzl, F.; Feinendegen, L.
    E.; Bremer, H. J. Lancet 1975, 1, 855.
53. Lombeck, I.; Schnippering, H. G.; Kasperek, K.; Ritzl, F.;
    Kostner, H.; Feinendegen, L. E.; Bremer, H. J. Zeit. Kinder-
    heilk. 1975, 120, 181-9.
54. Henkin, R. I.; Aamodt, R. L. Lancet 1975, 1, 1379-80.
55. Aamodt, R. L.; Rumble, W. F.; Johnston, G. S.; Henkin, R. I.
    Amer. J. Clin. Nutr. 1981, 34, 2648-52.

56. Nelder, K. H.; Hambidge, K.M. New Engl. J. Med. 1975, 292, 2648-52.
57. Dillaha, C.J.; Lorincz, A. L; Aavick, O. R. J. Amer. Med. Assoc. 1953, 152, 509-12.
58. Prasad, A. S.; Miale, A., Jr.; Farid, Z.; Sanstead, H.H.; Schulert, A. R.; Darby, W. J. Arch. Int. Med. 1963, 111, 407-28.
59. Henkin, R. I. Olfactory Rev. 1981.
60. Henkin, R. I."Looseleaf Series of Otolaryngology"; Harper and Row, New York, NY, 1982, p. 1-39.
61. Kosman, D. J.; Henkin, R. I. Amer. J. Clin. Nutr. 1981, 34, 118-19.
62. Henkin, R. I., Schechter, P. J.; Raff, M.S.; Bronzert, D.A.; Friedewald, W. T."Clinical Applications of Zinc Metabolism" C. C. Thomas, Springfield, IL, 1974, p 204-28.
63. Mallette, L. E.; Henkin, R. I. Amer. J. Med. Sci. 1976, 272, 164-74.
64. Henkin, R. I.; Mueller, C.; Wolfe, R. J. Lab. Clin. Med. 1975, 86, 175-80.
65. Henkin, R. I.; Lippoldt, R. E.; Bilstad J.; Edelhoch, H. Proc. Nat. Acad. Sci.1975, 72, 488-92.
66. Aamodt, R. L.; Rumble, W. F.; Johnston, G. W.; Foster, D.; Henkin, R. I. Amer. J. Clin. Nutr. 1979, 32, 559-69.
67. Foster, D. M.; Aamodt, R. L.; Henkin, R. I.; Berman, M. Amer. J. Phyisol., 1979, 237, R340-R349.
68. Aamodt, R. L.; Rumble, W. F.; Babcock, A. K.; Foster, D. M.; Henkin, R. I. Metabolism, 1982, 31, 326-34.
69. Babcock, A. K.; Henkin, R. I.; Aamodt, R. L.; Foster, D. M.; Berman, M. Metabolism 1982, 31, 335-47.
70. Shatzman, A. R.; Henkin, R. I. Proc. Nat. Acad. Sci. 1981, 78, 3867-71.

RECEIVED October 13, 1982

Utilization of Zinc by Humans

S. J. RITCHEY and L. JANETTE TAPER

Virginia Polytechnic Institute and State University, Department of Human
Nutrition and Foods, Blacksburg, VA 24061

The utilization and retention of zinc has
been studied in several groups of human subjects,
including preadolescent children, young adult
women, pregnant women, and elderly adults.
Studies, ranging in length from 18 to 42 days, were
conducted under controlled conditions, including
regulation of food and water intake, collection
of all excreta, and careful monitoring of activities of subjects. Experimental diets were
comprised of commonly consumed foods with supplements of zinc and/or other nutrients to achieve
the objective of the study or to obtain desired
intakes of critical nutrients. Apparent absorption of zinc ranged from about 2 percent in
elderly adults to about 60 percent in one group
of children. However, the usual range of apparent
absorption was between 20 and 30 percent of intake.
Average zinc excretion through feces, as a percentage of intake, ranged from 73.0 percent in
preadolescent children to 97.4 percent in the
elderly adults. Urinary zinc was between 0.3 and
0.4 mg/day for most subjects. In the elderly,
urinary excretion was about 0.8 mg/day. These
studies provide the basis for estimating losses
of zinc in relation to dietary intake and for
assessing physiological needs of human subjects.

During the past several years we have investigated the
utilization of zinc in several different age groups of human
subjects. Work with growing preadolescent children, young adults,
pregnant women, and elderly adults has been reported. Our goal
through this work has been to provide reasonable estimates of
zinc needs of the human at different ages and during normal
physiological events. This paper is not intended to be a compre-
hensive review, but focuses on work accomplished at laboratories
within this institution.

0097-6156/83/0210-0107$06.00/0

Experimental Approach

The balance study technique has been used in all of these studies. Although there are difficulties and certain problems with balance studies, nutrition researchers have not yet found a reasonable and realistic alternative to estimate nutrient needs of human subjects. Continued development and increased sophistication of methods and analytical techniques may provide alternatives in the future, but these are not clearly available at the present time.

Throughout our studies we employed the same general approach and techniques. We placed consideration emphasis on the enlistment of cooperative subjects who have an appreciation for the research and who understand the need for complete adherence to the guidelines of the specific study. In addition, we emphasized the need for strict control of all potential sources of contamination, an absolute essential for studies with the trace minerals. We were successful in both of these concerns with minor deviations.

The general approach to our studies has been described in detail (1). We utilized an adjustment period ranging from five to ten days dependent upon the specific experiment. Commonly consumed foods were used to develop a series of menus (Table I), usually three to six, to avoid boredom and because we wanted to have results as closely representative of the general population as feasible. Menus were developed and tested, key nutrients were analyzed for, and modifications were made as necessary prior to the beginning of an experiment. Supplements of nutrients were utilized to achieve the desired experimental intakes and/or to meet the daily recommended allowances of other essential nutrients (3) for the particular age group. Urine and fecal excretions were collected throughout the study. Appropriate composites of food and excreta were prepared and samples were taken for analytical work. Retention, apparent absorption, and other calculations were based on the entire study except the adjustment period, although in some studies the calculations were made for discrete periods within the experiment when there appeared to be changes in nutrient retentions as a function of an experimental treatment.

Dietary Protein and Zinc

Zinc received attention in our institution as early as 1956 in studies with preadolescent children. A series of experiments were done under completely controlled conditions (1) with the goal of defining nutrient needs of the growing child. Several investigators, representing agricultural experiment stations in the southern region and working under the regional research concept, participated in these studies. Engel et al. (4) and

Table I.   Example menu pattern used in
human studies.[a]

Breakfast

Orange juice	122	Apple juice	122
Bran flakes	28	Bran flakes	28
Toast	23	Hard cooked egg	44
Margarine	5	Toast	23
Coffee	2	Margarine	10
Milk	122	Coffee	2
Sugar	7	Milk	122
		Sugar	7

Lunch

Tuna salad	60	Cottage cheese	90
Lettuce	26	Tomato wedges	75
Crackers	11	Lettuce	26
Apple dumpling	75	Crackers	11
Punch	244	Cantaloupe	100
Mayonnaise	14	Lemonade	244

Dinner

Turkey breast	90	Baked fish	90
Potatoes	100	Baked potato	100
Green beans	100	Peas	50
Cranberry jelly	40	Carrots	50
Roll	26	Roll	26
Margarine	10	Margarine	10
Fruit cocktail	100	Sherbet	100
Coffee	2	Coffee	2
Sugar	7	Sugar	7
Punch, fruit	244	Lemonade	244
Graham crackers	28	Graham crackers	28

[a]List of menu items and weights (g). Analyzed values for key
nutrients were done prior to and throughout the experiment.
Zinc contents were 6.10 and 5.98 mg, respectively. The mean
protein content was 56 g and the mean energy content was 2000
kcals. Adapted from (2).

Price et al. (5) measured the retention of zinc and other minor
elements in studies in which the major variables were source and
level of dietary protein. Comparison of protein sources from
plant and mixed sources indicated that the apparent absorption
was somewhat lower from the diets with only plant proteins than
that from mixed sources. A lower level of protein (25 g/day)
caused a lower retention of zinc than the moderate protein
(46 g/day) diet, although conclusions were complicated because of
other variables.

In a recent study (6) with young adults, the effect of level
of dietary protein on zinc utilization was examined. Two levels
of protein, 49.2 and 94.9 g/day, and two levels of zinc, 9.78 and
19.20 mg/day, were variables in a 24-day study. Zinc retentions
were not significantly affected by the level of protein. However,
urinary excretion of zinc was significantly ($p < 0.05$) higher for
subjects consuming the higher protein diet. This was true only
during the last 6-day period and raises the question of appropriate length of human balance studies. The very high fecal
excretions for subjects on the higher protein, high zinc treatment may have been a oddity. These excretions resulted in a
negative net retention and may be misleading in our evaluation
of these data.

Reports from other laboratories have conflicted with regard
to the effects of protein on zinc utilization. Greger and
Snedeker (7) reported that apparent absorption of zinc was
affected by the level of dietary protein, phosphorus intake and
the interaction between protein and phosphorus. Apparent retention of zinc was highest with the high protein, moderate
phosphorus diet compared to other treatments, including low protein and either low or moderate phosphorus variables. Urinary
zinc was higher in subjects consuming the high protein diets.
Based on a lengthy metabolic study with men, Sandstead et al. (8)
have reported that the dietary requirement for zinc increased as
the level of protein intake increased. All of these investigations suggest that a higher amount of protein in the diet results
in poorer utilization and/or increased metabolic need for zinc.
The situation may prove to be similar to that for other minerals,
such as calcium.

Oral Contraceptive Use and Pregnancy in Adult Women

Alterations in blood levels of zinc during pregnancy and
during the use of oral contraceptive agents are well documented
(9, 10). However, the utilization and need for zinc during
pregnancy and when women are using hormonal contraceptive agents
have received little attention. Utilization of zinc during the
use of oral contraceptives was not affected (11). Contraceptive
users were compared to controls (non-users) in an 18-day balance
study. Retentions and fecal excretions were similar for the two
groups - both groups were in negative balance on a mean zinc

intake of 9.12 mg/day. Apparent absorptions could not be calculated since fecal excretions exceeded intake. Urine excretions
were not different for the treatments.

A recent study (12) focused on the zinc needs during the
second trimester of pregnancy. This study was different from
most in our laboratory because we permitted subjects to consume
their normal diets, thus the balance and excretion data represent
a range of intakes. We found zinc intakes falling at two levels,
6.7 and 22.0 mg/day. The higher intakes resulted from supplements
of zinc. We measured retentions in two weeks during the 4.5
month to the 5.5 month of pregnancy. The study design resulted
in varying levels of intake and retentions, particularly for the
lower, non-supplemented group of women. Both groups of subjects
were in net positive balance, although the lower level was marginal at best when all potential losses of zinc were accounted
for in estimating retentions. Apparent zinc absorption ranged
from 9.6 to 23.79 percent with no effect resulting from level of
dietary intake.

Taper et al. (13) investigated the effects of zinc intake on
copper utilization in adult women. The experiment involved feeding three levels of zinc, ranging from 8.0 to 24.0 mg/day. Copper
retention was not affected by zinc intake. Urine losses of zinc
were from 0.4 to 0.6 mg/day and fecal excretions paralleled the
intake. Percent apparent absorptions of zinc were 7.75, 9.13,
and 11.50 percent when intakes were 8.0, 16.0 and 24.0 mg/day.

Utilization in Growing Children

In addition to the study of Engel et al. (4) and Price et al.
(5), two other experiments have been conducted in our laboratories. Examining the effect of zinc supplementation on protein
utilization in the growing child, Meiners et al. (14) fed zinc
levels of either 5.5 mg/day, the amount present in diets typical
of low-income populations in the Southern region, or 10.5 mg/day.
Retentions were borderline equilibrium for subjects on the lower
intake, but were positive for those children consuming 10.5 mg/
day. Urine excretions were in the range of 0.45 to 0.79 mg/day.
Losses were higher than for most adult studies in our laboratory.
The apparent absorptions ranged from 7.1 to 21.2 percent.

Preadolescent girls were provided zinc levels ranging from
5.61 to 14.61 mg/day to evaluate the needs for zinc by this age
group (15). Percent apparent absorption ranged from 26.1 to 43.7
percent with most subjects in the more narrow range of 25 to 30
percent.

Losses of Zinc

The consideration of zinc utilization is in large measure an
assessment of zinc losses or excretion under any given set of
circumstances, including dietary variables, physiological state,

pathological insult, stress, and other conditions which may affect utilization. Major routes of zinc excretion are fecal and urinary, with potential losses through sweat, and probably some slight losses through other routes. Studies from our laboratories and from others are now such that we can begin to have some general notion of levels of excretion by human subjects and can predict in a general fashion the utilization of zinc.

Urinary losses of zinc represent small, and apparently fairly constant amounts. Using the data from several studies in our laboratory, we made calculations of average losses by groups of subjects (Table II). Urine excretion was usually in the range of 0.3 to 0.4 mg/day with considerable variability around those average values. However, we have a reasonable basis for suggesting that urinary losses will be consistently within the 0.3 to 0.5 mg/day range when subjects are consuming commonly used foods. From our data, elderly adults are the clear exception to this generalization, but there are limited numbers of individuals in this pool of data.

Fecal excretions were calculated as a percentage of intake since this loss almost always in our experience closely follows intake (Table II). Percentages of intake excreted through feces ranged from 73.0 in the preadolescent child to 97.4 in the elderly adults. Dietary factors may have more impact on the loss of zinc through the gastrointestinal tract, thus a wide range of excretions from study to study might be anticipated. It is interesting that the lowest percentages occur in those groups that would have the higher physiological needs for zinc, preadolescent children and pregnant women.

Losses of zinc through sweat can be significant, particularly in hot, humid climates. Data from a few studies, including those of Jacob et al. (16) and Ritchey et al. (15), suggest that losses through this route are in the range from 0.5 to 1.0 mg/day, perhaps somewhat higher in tropical areas.

Summary

In most of the individuals studied in our laboratory, zinc is absorbed within the range of 20 to 30 percent of intake, although there are significant deviations from that apparently normal response. Losses, including excretions and sweat loss, from commonly consumed diets can be predicted in a general way at the present time, thus utilization of zinc by human subjects can be estimated under most circumstances. Additional information and refinements are needed for some population groups, such as the elderly and pregnant women.

Table II. Excretions of zinc through urine and
feces by groups of subjects.

Group	No. of Subjects	Urine (mg/day) Mean	Urine (mg/day) Range	Feces (% of intake) Mean	Feces (% of intake) Range	Ref.
Preadolescent children	70	0.38	0.20-0.79	73.0	56.3-92.9	(15)
Adult females	57	0.37	0.27-0.57	93.2	82.3-104.7	(6, 11, 13)
Pregnant women	24	0.30	0.25-0.34	82.9	76.2-90.5	(12)
Elderly adults	10	0.78	0.46-1.09	97.4	96.3-98.5	(2)

Literature Cited

1. Ritchey, S. J.; Korslund, M. K. Home Econ. Res. J. 1976, 4,
    248–252.
2. Burke, D. M.; DeMicco, F. J.; Taper, L. J.; Ritchey, S. J.
    J. Gerontology 1981, 36, 558–563.
3. Food and Nutrition Board. "Recommended Dietary Allowances";
    National Academy of Sciences: Washington, 1980, 9th ed.
4. Engel, R. W.; Miller, R. F.; Price, N. O. In: Zinc
    Metabolism (Prasad, A, S., ed. C. C. Thomas, Springfield,
    Ill., 1956).
5. Price, N. O.; Bunce, G. E.; Engel, R. W. Am. J. Clin. Nutr.
    1970, 23, 258–260.
6. Colin, M. A. M.S. Thesis, Virginia Polytechnic Institute
    and State University, Blacksburg, VA, 1982.
7. Greger, J. L.; Snedeker, S. M. J. Nutr. 1980, 110, 2243–
    2253.
8. Sandstead, H. H.; Klevay, L.; Mahalko, J.; Johnson, L.;
    Milne, D. Am. J. Clin. Nutr. 1981, 34, 617 (Abs.).
9. Prasad, A. S.; Oberleas, D.; Lei, K. Y.; Moghissi, M. D.;
    Stryker, J. C. Am. J. Clin. Nutr. 1975, 20, 377–391.
10. Margen, S.; King, J. C. Am. J. Clin. Nutr. 1975, 28, 392–
    402.
11. Crews, M. G.; Taper, L. J.; Ritchey, S. J. Am. J. Clin.
    Nutr. 1980, 33, 1940–1945.
12. Oliva, J. T. M.S. Thesis, Virginia Polytechnic Institute
    and State University, Blacksburg, VA, 1981.
13. Taper, L. J.; Hinners, M. L.; Ritchey, S. J. Am. J. Clin.
    Nutr. 1980, 33, 1077–1082.
14. Meiners, C. R.; Taper, L. J.; Korslund, M. K.; Ritchey, S. J.
    Am. J. Clin. Nutr. 1977, 30, 879–882.
15. Ritchey, S. J.; Korslund, M. K.; Gilbert, L. M.; Fay, D. C.;
    Robinson, M. F. Am. J. Clin. Nutr. 1979, 32, 799–803.
16. Jacob, R. A.; Sandstead, H. H.; Munoz, J. M.; Klevay, L. M.;
    Milne, D. B. Am. J. Clin. Nutr. 1981, 34, 1379–1383.

RECEIVED October 6, 1982

Zinc Bioavailability from Vegetarian Diets

Influence of Dietary Fiber, Ascorbic Acid, and Past Dietary Practices

C. KIES, E. YOUNG, and L. McENDREE

University of Nebraska, Lincoln, NE 68583

Differences in food consumption patterns may result not only in changes in nutrient consumption patterns but also, directly or indirectly, in the availability of these nutrients. In a preliminary evaluation zinc utilization by omnivore (plant and animal product eaters) and vegetarian subjects consuming laboratory controlled vegetarian diets were compared. Fecal zinc excretions and zinc balances of subjects indicated better utilization of zinc from the vegetarian diets by the practicing vegetarian subjects than by the omnivore subjects consuming the vegetarian diets. While both fiber and ascorbic acid additions tended to decrease zinc utilization by omnivores and by vegetarians, the effect was less pronounced with the vegetarians than with the omnivores. This suggests that with time adaptation may have occurred. However, all subjects responded in a directionally similar manner to experimental variables of ascorbic acid and fiber supplementation.

Historically, the first human zinc-responsive syndrome was recognized in 1961 by Prasad (1,2) in Egyptian boys whose diets were composed largely of bread and beans. Zinc deficiency in these patients, characterized by dwarfism and hypogonadism, responded to zinc supplementation. Analysis of the food consumed by the human patients suggested it contained adequate zinc. Pigs also developed zinc deficiency symptoms when fed similar diets even though rations contained adequate zinc as judged by requirements set using purified rations (3). One factor common to both

0097-6156/83/0210-0115$06.00/0
© 1983 American Chemical Society

situations was the consumption of foods largely of plant origin,
cereals and legumes with little or no animal products. It has
generally been concluded that the deficiencies must have resulted
from a decreased availability of dietary zinc or an accelerated
loss of zinc from the body or both. These findings have implications for zinc nutrition of all classes of vegetarians.
Theoretically, as omnivore diets which contain variable amounts
of meat and vegetable products become less meat orientated and
more plant product orientated, the possibilities of inadequacies
in zinc nutrition become greater.

As reviewed recently by Solomons (4), many dietary and non-
dietary factors may affect the bioavailability of zinc. Dietary
factors are subdivided into "intrinsic factors" and "extrinsic
factors". "Intrinsic factors" relate to the chemical nature of
zinc itself. "Extrinsic factors" include non-heme iron, ethyl-
enediaminetetraacetic acid (EDTA), dietary fiber, phytic acid,
calcium, copper and specific foods such as cow's milk, cheese,
coffee, eggs, celery and lemon which have been demonstrated to
decrease zinc bioavailability as well as other factors which may
increase zinc utilization.

Generally foods of animal origin have been found to be higher
in zinc content than those of plant origin (5,7). Legumes, whole
grains, and leafy vegetables may contain significant amounts of
zinc but the zinc content of most fruits and vegetables is quite
modest (6). Processing of raw products also may have an adverse
effect on the zinc content of foods. White bread contains about
one-fourth of the zinc originally present in the unrefined wheat,
the zinc content of degermed corn meal is roughly one-half that
in dry corn, and precooked, quick polished rice (dry) contains
less than one-third the amount of zinc of unprocessed grain (5).

Zinc absorption from a number of food substances, particularly those of plant origins, has been questioned. Found widely
distributed in foods of plant origin, phytate has been demonstrated to be a chelating agent complexing with zinc to form a
compound insoluble at normal intestinal pH. Evidence indicates
that phytate impairs both the absorption of dietary zinc and the
reabsorption of endogenously secreted zinc (3,8,9). Excessive
dietary calcium in the presence of phytate has been found to have
particularly adverse effects on zinc absorption (8). Since high
levels of phytate and high levels of fiber tend to occur simultaneously in foods, the relative influence of these two factors
has been difficult to separate. Recently, it has been reported
that non-heme but not heme iron adversely affects zinc absorption
(10). Since fiber, phytate and non-heme iron are likely to be
supplied in larger amounts in vegetarian than in omnivore diets,
it would be assumed that the zinc content of vegetarian diets is
less well utilized than that of omnivore diets.

Zinc Nutritional Status of Vegetarians

Relatively few dietary surveys of nutrient intakes of vegetarians have included a calculation of zinc intakes. Handbook
analytical values for zinc contents of a wide variety of food
substances have only recently become available which may in part
account for this omission.

Zinc status of 49 Seventh Day Adventist Canadian women who
had been consuming vegetarian diets for an average of 19 ± 17
years were reported by Anderson et al. (11). As calculated from
3-day dietary diaries, mean dietary fiber intake of this group
was 30.9 ± 110 g/day, which is relatively high, and the mean zinc
intake was 9.2 ± 2.5 g/day, which is considerably less than the
NRC RDA recommendation of 15 mg/day. Furthermore, 82% of the
zinc intake came from products of plant origin including 26% from
dried legumes, nuts and soy products. Never-the-less, blood
serum zinc levels (99.3 ± 23.9 ug/dl) and hair zinc levels (187
± 44 ug/g) were well within normal levels. Thus, in spite of
low zinc intake with zinc supplied largely from foods believed to
contain zinc of low bioavailability, those women were seemingly
in adequate zinc nutritional status.

Zinc intakes of 6 Swedish vegans (individuals who consume no
food of animal origin at all) using chemical analyses of diets
from a duplicate portion sampling technique approach were included
in a study by Abdulla et al. (12). Dietary fiber intake of male
subjects was 62 ± 9 g/day and that of female subjects was 43 ± 9
g/day. Zinc intake of male subjects was 13 ± 2.3 mg/day and that
of female subjects was 6.5 ± 1.3 mg/day, so as in the study of
lacto-ovo-vegetarians by Anderson et al., subjects were found to
have relatively high intakes of fiber and relatively low intakes
of zinc with the zinc that was supplied being from foods believed
to exhibit low zinc bioavailability.

Zinc intakes and consumption of several other nutrients using
three-day diaries were determined for 12 pregnant ovo-lacto-vegetarian women, 5 non-pregnant ovo-lacto-vegetarian women, and 6
pregnant omnivore women by King et al. (13). These were compared
to several parameters of zinc nutritional status. Energy intake
of the non-pregnant vegetarian women was much lower for this
group than for the pregnant vegetarian or pregnant omnivore
groups (1329 vs. 2446 vs. 2003 kcal/day, respectively). Zinc
intakes from food for these three groups (non-pregnant vegetarian,
pregnant vegetarian and pregnant omnivores) were 6.4, 12.4, and
14.4 mg/day, respectively; and from food plus zinc supplements
were 6.4, 29.6 and 16.1 mg/day, respectively. Urinary zinc
excretion levels roughly followed the zinc intake pattern being
0.17, 0.34 and 0.41 mg/g creatinine, respectively, for the three
groups. Hair zinc levels were approximately the same for all
three groups. However, blood plasma zinc levels for the non-
pregnant vegetarians, the pregnant vegetarians and the pregnant
omnivores were 80, 60 and 66 ug/dl, respectively, suggesting that

this parameter of zinc status was more affected by pregnancy than
by level of zinc intake or by source of dietary zinc.

In a study conducted in this laboratory, 23 Seventh Day
Adventist students were subjects; 14 claimed to be vegetarians
and 9 claimed to be omnivores (14,15). All were students at
Union College-Lincoln, Nebraska, a Seventh-Day Adventist institution, and consumed most meals (and other food) at the Union
College cafeteria which is operated as a lacto-ovo-vegetarian
food service. Subjects consumed self-selected diets. Materials
collected included 3-day dietary diaries kept by the subjects
themselves, samples of all foods served in the food service unit
and serving sizes of all foods served, and fasting blood samples
during the study period from all subjects. Zinc contents of
food samples and blood serum samples were analyzed by atomic
absorption spectrophotometry using a Varian Techtron Atomic
Absorption Spectrophotometer.

Dietary zinc intake was calculated for each subject using
the 3-day dietary diary information and analytical data from food
sample analyses. Handbook information was used for foods not
consumed at the Union College cafeteria (5-7).

As shown in Table I, the mean zinc intakes of the vegetarian
subjects was found to be 8.49 mg/day ± 2.35 and that of the omnivores was 8.09 mg/day ± 2.30. Although the ranges in intake in
both groups were great, 4.78 - 14.89 mg for the vegetarians and
5.09 - 11.31 mg for the omnivore group, not a single subject had
a 3-day mean intake of zinc which met or exceeded 100% of the
NRC RDA's of 15 mg/day. Approximately one-half of the subjects
in both groups had 3-day mean values of less than 50% of the NRC
RDA's for zinc for the age/sex group investigated. Thus, it
would appear that some risk of zinc deficiency for these individuals might exist.

Mean blood serum zinc levels of the vegetarian subjects was
75.5 ug/dl ± 4.81 with a range of 67.9 - 86.0 ug/dl and that for
the omnivore subjects was 72.6 ug/dl ± 5.70 with a range of 64.8 -
83.1 ug/dl. Blood serum zinc levels were weakly positively correlated with intake levels but with the number of subjects
involved, this correlation was not statistically significant.
While the mean values were within normal ranges for blood serum
zinc levels, some individual values fell slightly below the 70
ug/dl lower acceptable level. Thus, blood serum zinc levels for
both groups could be said to be only marginally acceptable. The
omnivore subjects tended to have slightly lower blood serum zinc
levels just as they tended to have slightly lower intake levels
of zinc. It is important to restress that these individuals were
omnivores (self-designated) who actually were consuming very
little meat at the time the study was conducted.

It would appear on the basis of results of these four
studies that vegetarians tend to have low zinc intake levels.
Furthermore, most of the zinc in these diets appears to be provided from foods with low zinc bioavailability. Never-the-less,

zinc nutritional status as judged by biochemical parameters
appeared to be in the normal range.

Table I

Zinc Status of Vegetarians and Omnivores 7th Day Adventist
Students Eating in a Lacto-Ovo-Vegetarian Food Service

	Mean values	
	Vegetarians	Omnivores
No. of subjects	14	9
Zn intake, mg/day	8.49	8.09
Standard deviation	±2.35	±2.30
Range	4.78–14.89	5.09–11.31
% below 50% RDA	48	53
% above 75% RDA	26	9
% above 100% RDA	0	0
Blood serum Zn, ug/dl	75.5	72.6
Standard deviation	±4.81	±5.70
Range	67.9–86.0	64.8–83.1

Comparative Zinc Utilization by Vegetarians and Omnivores

Vegetarian diets tend to provide zinc in lower amounts than
that present in omnivore diets. Furthermore, zinc provided by
foods from vegetable origins tends to be less available than that
provided by foods of animal origin.

Effect of feeding a lacto-ovo-vegetarian diet on zinc status
was studies by Freeland et al. (16). Following feeding of a
lacto-ovo-vegetarian diet for three weeks, zinc absorption was
measured in 11 females. Zinc absorption was determined by zinc
tolerance tests in plasma and saliva after ingestion of 50 mg
zinc sulfate prior to and at the end of dietary treatment. The
zinc content of the subjects' saliva significantly decreased from
the initial average of 131 ppm to an average of 92 ppm. Plasma
levels showed a tendency to decrease. These changes indicate
that a vegetarian diet may reduce zinc absorption.

The objective of a project conducted in this laboratory was
to determine whether or not vegetarian subjects and omnivore
subjects utilize zinc to a similar degree when fed laboratory
controlled vegetarian diets (14,15).

This was a composited project involving the recombination of
data from several studies. Because of the considerable numbers
of Seventh-Day Adventists who reside in Lincoln, because of the
location of Union College, a Seventh-Day Adventist institution,

or because of interest of Seventh-Day Adventists in nutrition and
human health, we find that a number of vegetarians, primarily
Seventh-Day Adventists, are volunteering as diet subjects for our
general human nutrition metabolism studies. On the surface it
might seem advisable to bar participation of these individuals in
order to reduce variability within each experimental group population, but for ethical, legal and practical reasons this
approach was not really a viable option. In the current project,
data were drawn from several studies during which similar but not
identical vegetarian diets were fed for 21-28 days. While objectives of these studies varied and different experimental variables in the form of supplements were given, data were drawn
from control periods during which no supplements were given. Data
for 12 vegetarians and 12 omnivores were used. All foods fed
were sampled. Complete urine and stool collections were made.
Zinc content of urine, feces and food were determined by atomic
absorption spectrophotometry.

Mean zinc intakes varied slightly between the two groups
being 8.32 mg/day for the vegetarian group and 8.86 mg/day for
the omnivore group. Zinc intakes also varied somewhat among
subjects within each group. This variation was in part due to
variation in food intake because of differences in caloric intake
of the subjects and in part due to variation in zinc content of
the different basal diets used in the different studies. In any
case, zinc intake was marginally low in comparison to NRC RDA's
for zinc (15 mg/day) but, by chance, was reasonably close to
actual zinc intakes of Seventh-Day Adventist students consuming
self-selected diets as discussed earlier. Marginal intakes of
test nutrients are usually sensitive levels for biological evaluation. As shown in Table II, mean fecal zinc loss was 6.91 mg/
day for the vegetarian subjects and 9.10 mg/day for the omnivore
subjects, differences which were statistically significant
(P < 0.05). Urinary zinc losses of the two groups were not statistically different. Mean zinc balances of the vegetarian subjects
were +1.20 mg and were -0.47 mg/day for the omnivores. These
differences were statistically significant (P < 0.05) even though
considerable variation in individual subject values existed within
each group.

Various explanations for these results are possible.
Although we do not have data to substantiate the statement, it is
likely that the omnivores were receiving higher levels of zinc
prior to the start of the study than were the vegetarians, hence,
may have had the double problem of adjustment both to a vegetarian diet and a to a lower intake of zinc. Furthermore, since the
data were drawn from several studies rather than from one study,
the possibilities of spurious results or, worse, spurious conclusions are increased enormously.

In any case, it would appear that adaptation to vegetarian
diets may occur which enables vegetarians to make better use of
zinc in vegetarian diets than would be expected from results of

studies using omnivores consuming vegetarian diets for short periods of time. While the evidence presented here is only suggestive, not conclusive, we feel that it justifies further research effort. Because of this, comparative utilization of zinc by omnivores and vegetarians fed diets supplemented with fiber and with ascorbic acid was also examined.

Table II

Zinc Status of Vegetarian and Omnivore Subjects Fed
Laboratory Controlled Vegetarian Diets

| | Mean values | |
	Vegetarians	Omnivores
No. of subjects	12	12
Zn intake, mg/day	8.32	8.86
Fecal Zn, mg/day	6.91[a]	9.10[b]
Urine Zn, mg/day	0.21[a]	0.23[b]
Zn balance, mg/day	+1.20[a]	-0.47[b]

[ab] Values with different letter superscripts are significantly different from one another at P 0.05.

Effect of Dietary Fiber on Zinc Utilization by Vegetarians and Omnivores

Increasing dietary fiber from a wide variety of sources in laboratory controlled human feeding studies has been found to increase fecal zinc excretion and, presumably, to decrease absorption of this nutrient (17-28). Dietary fiber may adversely affect the absorption of zinc by diluting the concentration of zinc in the intestinal chyme, by decreasing time for absorption by decreasing fecal transit time, by trapping zinc within particles containing fiber, and/or by providing surfaces for zinc absorption and thus hindering absorption through the intestine.

In our laboratory, the comparative effects of supplementation of a laboratory controlled vegetarian diet with three different fiber sources was examined when fed to omnivore and vegetarian subjects (14,15).

Six subjects who were participants in a study on effects of several purified fibers on nutrient utilization and who were vegetarians were age, sex, height, weight and ethnic-group matched with six subjects from this series of studies who were omnivores. The laboratory-controlled diet fed to all subjects for 21-28 days was a lacto-vegetarian type. The basal diet provided 10.10 mg of zinc and 14.7 g of fiber. Cellulose, hemicellulose or pectin were added to the basal diet to provide 20 g fiber/day in separate, randomly arranged period of 6 to 7 days each.

Results are shown in Table III. Urinary zinc excretion of
omnivore and vegetarian subjects were not significantly different
from one another whether the laboratory controlled vegetarian
diet was fed alone, or with fiber supplements. Fiber supplementation did not significantly affect the urinary excretion of
zinc in either group. Fecal zinc excretions of omnivore subjects
were significantly greater than were those of vegetarian subjects
regardless of whether or not fiber supplements were added to the
lacto-vegetarian diet. Hemicellulose supplementation resulted
in a significant increase in fecal zinc loss with both omnivore
and vegetarian subjects; however, the degree of increased loss
was greater for omnivore than for the vegetarian subjects. Zinc
balances for the omnivore subjects fed the lacto-vegetarian
diet were negative whether or not fiber supplements were employed
while those of the vegetarian subjects were in the positive
range. Addition of hemicellulose and cellulose had a depressing
effect on zinc balances of both groups while pectin exerted
relatively little effect.

Table III

Effect of Fiber Supplementation on Zinc Status of Vegetarian
and Omnivore Subjects Fed a Laboratory Controlled Vegetarian Diet

Parameter	Diet	Vegetarians	Omnivores
Urinary zinc	Basal	0.27^a	0.31^a
(mg/day)	Basal + 14.7 g cellulose	0.22^a	0.22^a
	Basal + 14.7 g hemicellulose	0.21^a	0.35^a
	Basal + 14.7 g pectin	0.25^a	0.37^a
Fecal zinc	Basal	9.38^a	10.33^c
(mg/day)	Basal + 14.7 g cellulose	9.66^{ab}	11.02^d
	Basal + 14.7 g hemicellulose	9.72^b	11.87^d
	Basal + 14.7 g pectin	9.40^a	10.42^c
Zinc balance	Basal	$+0.55^a$	-0.54^c
(mg/day)	Basal + 14.7 g cellulose	$+0.22^{ab}$	-1.14^d
	Basal + 14.7 g hemicellulose	$+0.17^b$	-2.12^d
	Basal + 14.7 g pectin	$+0.55^a$	-0.69^c

Values with different letter superscripts are significantly
different from one another at P <0.05.

Influence of Ascorbic Acid Supplementation on Zinc Utilization by Vegetarians and Omnivores

Ascorbic acid is known to increase the intestinal absorption
of non-heme iron but to decrease the absorption of copper (29-35).

Since ascorbic acid is a strong reducing agent, it is presumed to reduce ferric iron to its ferrous form or to maintain ferrous iron in its reduced state which is more available. However, its action on copper would be to reduce Cu (III) to Cu (I), the less available form of copper (36). Ascorbic acid is also thought to increase the solubility of iron by decreasing the alkalinity of the intestinal chyme. Effect of ascorbic acid on the availability of zinc has been less extensively studied.

Theoretically, ascorbic acid might affect the utilization of zinc either by change in valence state or indirectly by increasing iron absorption and thus decreasing absorption binding sites for zinc. Patterns of zinc absorption were not significantly changed as a result of feeding doses to humans representing a spectrum of zinc to ascorbic acid molar ratios in a study conducted by Solomons et al. (36). These authors postulated that the lack of response was due to the resistance of zinc to undergo oxidation or reduction due to a filled d^{10} orbital in its third electron shell. However, in human balance studies conducted in this laboratory, ascorbic acid tended to increase fecal zinc losses and to decrease zinc balances (33,37). Single dose studies and longer term balance studies are different approaches to studying nutrient utilization which do not necessarily yield similar results.

In the forementioned studies (33,37), four vegetarian subjects were included. These were age, sex, race, height, and weight matched with four omnivore subjects from the same studies. The usual mixed food, laboratory controlled diet used in this series of studies did contain both ground beef and tuna fish. For the vegetarian subjects, an extruded, defatted soymeal product processed to resemble ground beef was used instead of the meat and fish products. Hence, the vegetarian subjects received a vegetarian diet closely matching the omnivore diet which was eaten by the omnivore subjects. The ascorbic acid content of both basal diets was 52 mg/subject/day. The zinc content of the vegetarian diets was 9.61 mg/subject/day and that of the omnivore diet was 10.53 mg/subject/day. During one period the basal diets were fed alone and in another period an ascorbic acid supplement of 200 mg/day was fed.

As shown on Table IV, mean serum zinc levels of vegetarian and omnivore subjects were similar while subjects received the basal diets without ascorbic acid supplementation. Ascorbic acid supplementation had no effect on serum zinc levels of the vegetarian subjects; however, mean serum zinc levels of omnivore subjects were slightly (but not significantly) depressed by the same treatment.

Mean urinary zinc excretion of omnivore and vegetarian subjects were not significantly different from one another nor did the addition of ascorbic acid supplements seem to have a pronounced effect (Table IV). Fecal zinc losses were significantly higer for the omnivore subjects than for the vegetarian

subjects even taking into consideration that the vegetarian diet
was approximately a milligram lower in zinc content than was the
omnivore diet. Addition of ascorbic acid increased fecal zinc
loss of both vegetarian subjects and omnivore subjects but the
increase in loss was considerably greater in the case of the
omnivore subjects.

These differences in mean fecal zinc losses are reflected
in the zinc balances (Table IV). It is interesting to note that
even though the omnivore subjects were receiving an omnivore diet
containing meat and fish, the mean zinc balance was negative
while that of the vegetarian subjects was positive. It may be
that the vegetarian diet more closely resembled the usual diet
of the vegetarian subjects than did the omnivore diet in the
case of the omnivore subjects, who probably were used to consuming diets containing much more meat providing larger amounts
of zinc. In any case, directionally, ascorbic acid supplementation tended to decrease zinc utilization of both groups of
subjects.

Table IV

Effect of Ascorbic Acid (AA) on Zinc (Zn) Utilization From
Omnivore and Vegetarian Diets Fed to Omnivore and Vegetarian
Subjects

Parameter	Vegetarians[1]		Omnivores[2]	
	−AA	+AA	−AA	+AA
Zn intake (mg/day)	9.61	9.61	10.53	10.53
Ascorbic acid intake (mg/day)	52	252	52	252
Blood serum Zn (ug/dl)	70^a	70^a	70^a	68^a
Urinary Zn (mg/day)	0.31^a	0.30^a	0.35^a	0.35^a
Fecal Zinc (mg/day)	9.05^a	9.22^b	12.04^c	12.48^d
Zn balances (mg/day)	$+0.25^a$	$+0.09^a$	-1.86^c	-2.30^d

[1] 4 vegetarian subjects fed a laboratory controlled vegetarian diet

[2] 4 omnivore subjects fed a laboratory controlled omnivore diet
Values with different letter superscripts are significantly
different from one another at P <0.05.

Conclusion

Zinc content of vegetarian diets tended to be lower than omnivore diets. Phytates, non-heme iron, fiber and ascorbic acid may inhibit utilization of zinc. These also tend to be found in larger amounts in vegetarian than in omnivore diets. Even so, biochemical indices of zinc nutritional status from survey studies have not identified vegetarians as being in poor zinc nutritional status. In laboratory controlled studies, omnivore subjects tend to utilize zinc from vegetarian diets less efficiently than do vegetarians. Solomons (10) and Cousins et al. (38) have suggested that regulation of zinc uptake is in accordance with nutritional requirements of the host and is governed by a feedback regulation mechanism. Thus, in short term laboratory controlled studies such as those given in the current project, it is likely that omnivore subjects had insufficient adjustment time. If this is true, then results of dose-type studies designed to study zinc absorption raise even more questions. It is of interest to note, however, that omnivores and vegetarians responded in similar directions to the variables of ascorbic acid intake and fiber intake even though the total amount of zinc absorbed varied between the groups.

Acknowledgments

Supported by Nebraska Agriculture Experiment Station Project 91-024 and USDA CSRS Project W-143. Published as Nebr. Agric. Expt. Station Journal Article Series 6869 .

Literature Cited

1. Prasad, A.S., Miale, Jr., A., Farid, Z., Sandstead, H.H., Schubert, A.R. and Darby, W.J. J. Lab. Clin. Med. 1963a, 61, 537.
2. Prasad, A.S., Schubert, A.R., Miale, Jr., A., Farid, Z. and Sandstead, H.H. J. Lab. Clin. Med. 1963b, 61, 537.
3. O'Dell, B.L. Am. J. Clin. Nutr. 1969, 22, 1250.
4. Solomons, N.W. J. Amer. Dietet. Assoc. 1982, 80, 115.
5. Murphy, E., Willis, B. and Watt, B.K. J. Amer. Dietet. Assoc. 1975, 66, 345.
6. Haeflein, K.A. and Rasmussen, A.I. J. Amer. Dietet. Assoc. 1977, 70, 610.
7. Freeland, J.H. and Cousins, R.J. J. Amer. Dietet. Assoc. 1976, 68, 526.
8. O'Dell, B. and Savage, J.E. Proc. Soc. Expt. Biol. Med. 1960, 103, 304.
9. O'Dell, B.L., Burpo, C.E. and Savage, J.E. J. Nutr. 1972, 102, 653.
10. Solomons, N.W. and Jacob, R.A. Am. J. Clin. Nutr. 1981, 34, 475.

11. Anderson, B.M., Gibson, R.S. and Sabry, J.H. Am. J. Clin. Nutr. 1981, 34, 1042.
12. Abdulla, M., Anderson, I., Asp, N.G., Berthelsen, K., Birkhed, D., Denckir, I., Johansson, C.G., Jagerstad, M., Kolar, K., Nair, B.M., Nilsson-Ehle, P., Norden, A., Rassner, S., Akesson, B. and Ockerman, P.A. Am. J. Clin. Nutr. 1981, 34, 2464.
13. King, J.C., Stein, T. and Doyle, M. Am. J. Clin. Nutr. 1981, 34, 1049.
14. Kies, C., Young, E., McEndree, L., Lo, B.F., Beshgetoor, D., and Fox, H.M. FASEB annual meeting (abstract no. 50263), 1981.
15. Young, E., M.S. Thesis-University of Nebraska, 1980.
16. Freeland, J.H., Ebangit, M.L. and Johnson, P. Fed. Proc. 1978, 37, 253.
17. Reinhold, J.G., Ismail-Beigi, F. and Faradji, B. Nutr. Rep. Inter. 1975, 12, 75.
18. Reinhold, J.G., Faradji, B., Abadi, P., and Ismail-Beigi, F. "Trace Elements in Human Health and Disease"; Academic Press: New York (Prasad, A.S., ed.), 1976.
19. Reinhold, J.G., Faradji, B., Abadi, P. and Ismail-Beigi, F. J. Nutr. 1976, 106, 493.
20. Sandstead, H.H., Munoz, J.M., Jacob, R.A., Klevay, L.M., Reck, S.J., Logan, G., Dintzis, F.R., Inglett, F.E. and Shuey, W.C. Am. J. Clin. Nutr. 1978, 31, S180.
21. Ismail-Beigi, F., Reinhold, J.G., Faraji, B. and Abadi, P. J. Nutr. 1977, 107, 510.
22. Ismail-Beigi, F., Faradji, B. and Reinhold, J.G. Am. J. Clin. Nutr. 1977, 30, 1721.
23. Lo, B.F. M.S. Thesis-University of Nebraska, 1979.
24. Papakyrikos, H. M.S. Thesis-University of Nebraska, 1979.
25. Kies, C. and Fox, H.M. Cereal Foods World, 1976, 8, 453.
26. Drews, L., Kies, C. and Fox, H.M. Am. J. Clin. Nutr. 1979, 32, 1893.
27. Kies, C., Beshgetoor, D. and Fox, H.M. "Antinutrients and Natural Toxicants in Foods", Food and Nutrition Press: Westport, CT (Ory, R.L., ed.), 1980, p. 319.
28. Beshgetoor, D. M.S. Thesis-University of Nebraska, 1977.
29. VanCampen, D., and Gross, E. J. Nutr. 1968, 95, 617.
30. Cook, J.D. and Monsen, E.R. Fed. Proc. 1977, 36, 2028.
31. Bjorn-Rasmussen, E. and Hallberg, L. Nutr. Metab. 1974, 16, 94.
32. Ghani, N.A. M.S. Thesis-University of Nebraska, 1981.
33. Lorenz, J.A. M.S. Thesis-University of Nebraska, 1981.
34. Bylund, D.M. M.S. Thesis-University of Nebraska, 1979.
35. Barmann, R.M. M.S. Thesis-University of Nebraska, 1981.
36. Solomons, N.W., Jacob, R.A., Pineda, O. and Viteri, F.E. Am. J. Clin. Nutr. 1979, 32, 2495.
37. Sui, F. M.S. Thesis-University of Nebraska, 1981.
38. Cousins, R.J. Am. J. Clin. Nutr. 1979, 32, 339.

RECEIVED October 13, 1982

Effect of Fiber and Oxalic Acid on Zinc Balance of Adult Humans

JUNE L. KELSAY

U.S. Department of Agriculture, Carbohydrate Nutrition Laboratory,
Beltsville Human Nutrition Research Center, Agricultural Research Service,
Beltsville, MD 20705

In a study of 12 men fed a diet containing fiber in
fruits and vegetables, mean zinc balance was negative
and significantly lower than that on a low fiber
diet. The higher fiber diet included spinach, which
is high in oxalic acid. In a second study, zinc
balance was not negative when cauliflower, which is
low in oxalic acid, replaced the spinach in the
higher fiber diet. In a third study, zinc balance
was negative and significantly lower on the higher
fiber diet including spinach than on a low fiber
diet including spinach, or on a higher fiber diet
without spinach. The adverse effect of the higher
fiber diet containing spinach was observed during
the fourth week of the study, but not during the
third week.

Fiber components can bind zinc and other minerals, possibly
rendering the minerals unavailable for absorption by the animal
body ($\underline{1}$, $\underline{2}$, $\underline{3}$). The effect of fiber on zinc balance of human
subjects was reviewed previously ($\underline{4}$) and appears to be related to
level and kind of fiber, level of zinc, other components of the
diet, and length of study period. One of the dietary components
which may affect zinc availability is oxalic acid.

Oxalic acid is capable of combining with minerals to form
salt complexes. Calcium and zinc form the least soluble salts
with oxalic acid ($\underline{5}$, $\underline{6}$). Spinach contains more oxalic acid than
most foods (approximately 700 mg/100 g), and its effect on calcium
availability has been studied rather extensively. Other green
leafy vegetables also contain considerable amounts of oxalic acid,
and the oxalic acid is concentrated more in the leaves than in the
stalks ($\underline{7}$). Rhubarb, some nuts, tea, and cocoa have also been
found to contain oxalic acid in amounts greater than 200 mg/100 g
food ($\underline{7}$-$\underline{11}$).

A number of studies have been conducted on rats to evaluate

the effect of the inclusion of spinach in the diet on calcium
utilization (12). Many of these studies were conducted with low
calcium intakes and high levels of spinach in the diet. Under
these conditions, when young rats were fed spinach, growth and
calcium in bones decreased. Calcium retention could be improved
by adding extra calcium, and was also influenced by the
availability of vitamin D.

Studies on the effect of oxalic acid in spinach on calcium
balance in humans have shown a small decrease or no effect on
calcium balance (12). However, when subjects were given test
meals of either Swiss chard (13) or amaranth (14, 15), which are
rich in oxalic acid, urinary excretion of calcium indicated that
the absorption of calcium from these sources was less than that of
an equal amount of calcium from milk. Absorption of calcium from
milk was also reduced when given along with amaranth (14).

Little information is available on the effect of oxalic acid
on zinc bioavailability. In one study, Welch et al. (16) fed
weanling rats zinc-deficient diets with and without 0.75% sodium
oxalate. The rats were dosed orally with zinc-labeled spinach
leaves or zinc-labeled zinc sulfate. Dietary oxalate enhanced the
availability of radioactive zinc from zinc sulfate, but had no
effect on zinc from spinach leaves. Absorption and retention of
zinc was greater from spinach leaves than from zinc sulfate.

Experiments

Three studies were carried out in our laboratory in cooperation with the Food, Nutrition, and Institution Administration
Department of the University of Maryland. The purpose of these
studies was to determine mineral balances in adult human subjects
when fiber in fruits and vegetables was added to the diet. Zinc
was one of the minerals determined. The effects of the diet
containing fiber were compared with those of a low fiber diet in
which fruit and vegetable juices replaced the fruits and vegetables. In the first study, spinach was included in the diet
containing fiber in fruits and vegetables. In the second study,
cauliflower, which is low in oxalic acid, replaced the spinach.
In the third study, spinach was included in the low fiber diet and
in a diet containing fruits and vegetables; cauliflower was
included in a second diet containing fruits and vegetables.

Study 1. Two diets were fed in a crossover design to 12 men
37 to 58 years of age (17). During the first 26 days of the
study, six men consumed the higher fiber diet containing fruits
and vegetables and six men consumed the low fiber diet containing
fruit and vegetable juices. At the end of 26 days, the subjects
consumed the alternate diet for 26 days. A 2-day revolving menu
comprised each diet. Foods included in the same amounts in both
diets are given in Table I. In Table II are listed the foods and
amounts fed in each of the two diets in addition to those in

Table I
Foods Common to Both Diets, 2800-Calorie Level, Study 1

Day 1	Day 2
Puffed rice cereal, 7.5 g	
Milk, 366 g	Milk, 366 g
Bread, white, 78 g	Bread, white, 103 g
Egg, scrambled, 50 g	Bacon, broiled, 15 g
Butter, 42.3 g	Butter, 28.2 g
Cream, 90 g	Cream, 90 g
Sugar, 60 g	Sugar, 72 g
Roast beef, 85 g	Tunafish, 78 g
Pudding, 130 g	Mayonnaise, 28 g
Ham, 85 g	Ground beef, 113 g
Salad dressing, 48 g	Salad dressing, 32 g
Cake, 53 g	Ice cream, 66 g

Table II
Additions to Diets, 2800-Calorie Level, Study 1

Day 1		Day 2	
Low fiber	Higher fiber	Low fiber	Higher fiber
Grapefruit juice, 247 g	Grapefruit, 200 g	Orange juice, 249 g	Oranges, 200 g
Jelly, 25 g	Dates, 80 g	Jelly, 25 g	Raisins, 72 g
Sugar, 24 g		Sugar, 18 g	
	Corn, 82 g		Peas, 85 g
Pineapple juice, 125 g	Pineapple, 123 g	Apple juice, 124 g	Apple, 160 g
Vegetable juice, 121 g	Spinach, 102 g	Vegetable juice, 121 g	Broccoli, 92 g
Macaroni, 105 g	Carrots, 55 g	Macaroni, 140 g	Squash, 120 g
	Cabbage, 45 g		Lettuce, 30 g
			Tomato, 100 g
Grape juice, 253 g	Blackberries, 252 g	Grape juice, 253 g	Blueberries, 284 g
Milk, 122 g		Milk, 122 g	

Table I. These are the amounts fed at the 2800-calorie level. Caloric intake was adjusted to the individual needs of the subjects by increasing or decreasing all foods the appropriate percentage to maintain each subject's weight during the study. Carbohydrate, fat, and protein made up 50, 37, and 13% of the total calories, respectively. The low fiber diet was supplemented with carotene, iron, magnesium, and copper in an attempt to make the two diets equivalent in all respects except fiber. During the last 7 days of each 26-day period, the subjects collected all urine and fecal samples. Feces were marked by giving each subject 50 mg Brilliant blue. Seven-day composites of food, urine, and feces were prepared. Neutral detergent fiber (NDF) intake was 5 and 25 g/day on the low and higher fiber diets, respectively (18). Bowel transit time was computed as the time from ingestion of the marker until most of the color appeared in the feces. Transit time decreased from a mean of 52 hours on the low fiber diet to a mean of 38 hours on the higher fiber diet. The inclusion of fruits and vegetables in the diet also increased fecal weight, number of defecations, and fecal excretions of energy, nitrogen, and fat.

Zinc intakes of the subjects were 13.2 and 12.6 mg/day on the low and higher fiber diets, respectively (19). Fecal excretions of zinc were 70 and 102% of the intake on the low and higher fiber diets, respectively (Figure 1). The mean zinc balance was positive on the low fiber diet and negative on the higher fiber diet, and these balances differed significantly from each other. The mean retention of zinc on the low fiber diet was 3.5 mg/day, and the mean loss on the higher fiber diet was 0.9 mg/day (Figure 2). These results indicated that increasing the fiber in the diet might result in negative balances of zinc, and appeared to confirm results of some earlier studies which were reviewed previously (4).

Study 2. The purpose of this study was to determine the effect of different levels of fiber in the diet on mineral balances. Four diets were fed to 12 men 35 to 49 years of age (20). All subjects consumed all diets for 21 days each in a 4 X 4 Latin square design. Diets 1 and 3 were similar to the low and higher fiber diets of study 1, except that a third menu, designed to contain the same amounts of nutrients and fiber as those in study 1, was rotated with the other two. Also, cauliflower replaced the spinach in the day 1 menu for diet 3. Diet 2 contained 1/2 and diet 4 contained 1 1/2 the amounts of fruits and vegetables of diet 3, and these two diets also had 3-day revolving menus. As an example, the foods and amounts fed at the 2800-calorie level for the day 3 menu are listed in Tables III and IV. Menus for days 1 and 2 for diets 2 and 4 were adjusted for intakes of fruits and vegetables in a manner similar to those for day 3. Caloric intakes were adjusted by increasing or decreasing all foods the appropriate percentage to maintain each subject's weight during

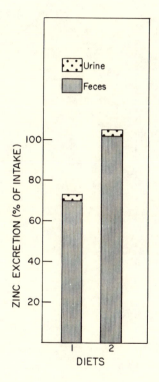

Figure 1.   Zinc excretion as percentage of intake, study 1 (19).

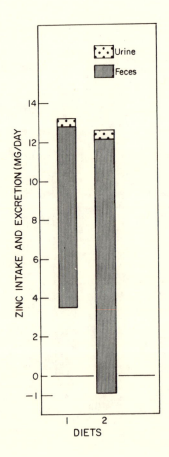

Figure 2. Zinc intake and excretion, study 1. Dietary intake is indicated by the top of the bar. The length of the vertical bar represents total excretion in urine and feces. Balance is indicated by the bottom of the bar. When the bottom of the bar is above the zero line, the balance is positive. When the bottom of the bar is below the zero line, the balance is negative.

Table III

Foods Common to All Diets, 2800-Calorie Level, Day 3, Study 2

Egg, 50 g	Bread, white, 50 g	Butter, 52 g
Cream, 90 g	Pork chop, 85 g	Mayonnaise, 14 g
Ice cream, 66 g	Turkey, 75 g	Cookies, 25 g

Table IV

Additions to Diets, 2800-Calorie Level, Day 3, Study 2

	Diet 1	Diet 2	Diet 3	Diet 4
	g	g	g	g
Orange juice	249	124		
Oranges		100	200	300
Figs, dried		38	75	113
Bread	28	28	28	
Maple sirup	39	39	39	
Milk	366	305	122	
Sugar	84	96	36	
Apple juice	124	62		
Pears		78	155	233
Brussels sprouts		39	78	116
Vegetable juice	121			
Asparagus		48	95	143
Rice	124			
Cranberry juice	100			
Cranberries		24	48	72
Apples		25	50	75
Raisins		25	50	75
Butterscotch pudding	130	65		
Pumpkin pudding		91	182	273
Cookies			10	

the study. Carbohydrate, fat, and protein made up 50, 37, and 13%
of the total calories, respectively, and were similar on all four
diets. Carotene, iron, magnesium, and copper were not added to
the low fiber diet as in study 1, because we felt that the unsup-
plemented intakes would be more representative of the levels of
intake on a low fiber diet.
     The number of defecations increased as the fiber in the diet
increased. The increase was from 7 per week on the low fiber diet
to 11 per week on the highest fiber diet; however, mean transit
time on all diets was about 30 hours. Both wet and dry fecal
weights increased as the fiber in the diet increased. Fecal
excretions of energy, nitrogen, and fat increased as the level of
fiber in the diet increased.
     Mean zinc intakes on the four diets ranged from 11.8 to 12.4
mg/day. Fecal excretions of zinc were 73, 74, 81, and 84% of the
intake on diets 1 to 4, respectively, and the percent excretion on
diet 4 was significantly higher than those on diets 1 and 2
(Figure 3). Zinc balance was significantly lower on diet 4 than
on diets 1 and 2, and retentions on the four diets were 2.8, 2.9,
2.1, and 1.5 mg/day for diets 1 to 4, respectively (Figure 4).
     The reason for the differences in zinc balances between these
two studies was not clear. In an effort to determine if the
oxalic acid in spinach in the first study could be responsible for
at least part of the decrease in bioavailability of zinc in the
first study, we determined oxalic acid content of feces. Mean
oxalic acid excretion in feces when the subjects were on the
highest fiber diet in study 2 was about 1/2 that on the higher
fiber diet in study 1 (210 and 423 mg/day, respectively). Another
factor which might help explain the results was the length of the
dietary periods; the dietary periods in study 1 were 5 days longer
than in study 2.

     Study 3. Because of the conflicting results of the first two
studies with regard to mineral balances, we carried out a third
study to investigate the separate and combined effects of fiber
and oxalic acid, as well as length of study period on mineral
balances.
     Twelve men 34 to 58 years of age consumed three diets for 28
days each in a 3 X 3 Latin square design (21). Diet 1 was similar
to the low fiber diet fed in the first study, with the following
modifications: Spinach was included in the diet every other day
and broccoli was included on alternate days; only 1/2 cup of milk
was given with the evening meal in place of 1 cup as in the first
study, in order to equalize the amount of calcium available with
the spinach meal in diets 1 and 2; carotene, iron, magnesium, and
copper were not added to the diet as in the first study. Diet 2
contained fiber in fruits and vegetables and was the same as the
higher fiber diet in the first study, including spinach every
other day and broccoli on alternate days. Diet 3 was the same as
diet 2 except that cauliflower replaced the spinach. On all three

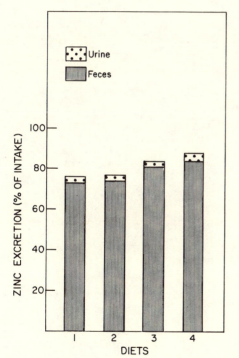

Figure 3.   Zinc excretion as percentage of intake, study 2 (20).

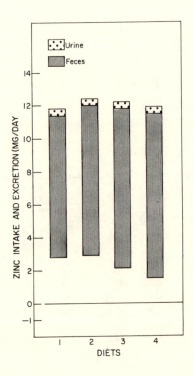

Figure 4. Zinc intake and excretion, study 2. Dietary intake is indicated by the top of the bar. Balance is indicated by the bottom of the bar.

diets, 1 cup of milk was given with breakfast and 1/2 cup with
dinner. During weeks 3 and 4 of each 28-day period, all urine and
feces were collected. Two 7-day composites of food, urine, and
feces were prepared.

Bowel transit times were not significantly different on the
three diets during week 3, but the mean transit times of 33 and 30
hours on the two higher fiber diets were significantly lower than
that of 43 hours on the low fiber diet during week 4. The number
of defecations and wet and dry fecal weights were significantly
higher on the diets containing fiber than on the low fiber diet
during both weeks 3 and 4.

Mean zinc intakes ranged from 10.7 to 12.3 mg/day. Fecal
zinc as a percent of intake on diets 1, 2, and 3 during week 3 was
95, 83, and 88%, and these values were not significantly different
from each other (Figure 5). Zinc balances during week 3 were 0.0,
1.7, and 0.7 mg/day on diets 1, 2, and 3, respectively, and were
not significantly different from each other (Figure 6). During
week 4, fecal zinc as a percent of intake was 77, 105, and 91% on
diets 1, 2, and 3, respectively, and was significantly higher on
diet 2 than on diets 1 and 3 (Figure 7). Zinc balances during
week 4 were 2.0, -1.4, and 0.4 mg/day on diets 1, 2, and 3,
respectively, and the balance on diet 2 was significantly lower
than those on diets 1 and 3 (Figure 8).

Discussion and Conclusions

Results of the third study indicate that the negative zinc
balances were a result of the combination of oxalic acid and fiber
in the same diet. With the level and kind of fiber and the level
of zinc fed, fiber alone did not have an adverse effect on zinc
balance. There was a decrease in zinc balance on the highest
fiber diet in study 2; however, the balance was positive. With
the level of zinc fed, a greater effect might have been seen with
levels of fiber higher than the ones fed. Some of the subjects
complained of the larger amounts of watery feces when they consumed the highest fiber diet in study 2. However, with an
extended period of time on this diet they might have become
adjusted and excreted stools with a firmer consistency. Also, an
extended period of time might have resulted in further decreases
in zinc balance. Bowel transit times, number of defecations,
fecal weights, and fecal excretions of energy, nitrogen, and fat
were not related to zinc balances, but were related to level of
fiber intake.

Further investigation is needed on the combined effects of
oxalic acid and fiber. In vitro studies could provide clues to
the nature of the binding of minerals to oxalic acid and fiber:
whether oxalic acid and fiber each binds part of the zinc, or
whether there is a fiber-zinc-oxalate complex formed.

Kojima et al. (22) studied the in vitro effect of spinach on
the solubilization of iron previously solubilized from cooked

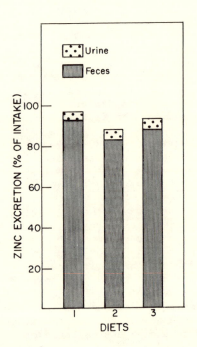

Figure 5.   Zinc excretion as percentage of intake, week 3, study 3.

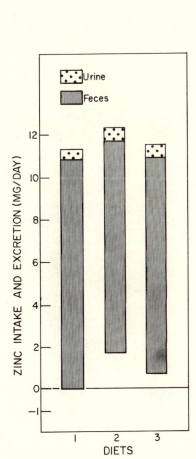

Figure 6. Zinc intake and excretion, week 3, study 3. Dietary intake is indicated by the top of the bar. Balance is indicated by the bottom of the bar.

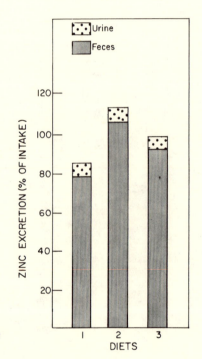

*Figure 7.   Zinc excretion as percentage
of intake, week 4, study 3.*

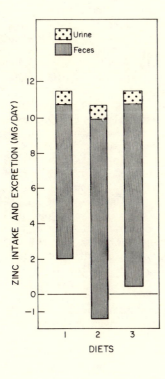

Figure 8. Zinc intake and excretion, week 4, study 3. Dietary intake is indicated by the top of the bar. Balance is indicated by the bottom of the bar.

pinto bean suspension. Total soluble iron was decreased by 7%
when spinach supernatant was added, by 50% when spinach residue
was added, and by 80% when whole spinach was added. The investigators concluded that "spinach supernatant contains factors which
bind iron and increase the affinity for association with the
insoluble spinach residue". At least one of these factors could
be oxalic acid. However, Van Campen and Welch (23) reported that
in rats iron was equally available from spinach, iron oxalate, and
$FeCl_3$, and that the addition of 0.75% oxalate to the diet appeared
to enhance iron utilization.

Since we did not find adverse effects on zinc balance by the
higher fiber diet containing spinach until the fourth week of
study 3, balance studies should be carried out for longer periods
of time to determine if negative zinc balances persist. Perhaps
in the earlier studies on effects of spinach on calcium balance of
human subjects, more definitive effects would have been noted if
the spinach had been fed along with a higher fiber diet. Length
of study period is also a likely factor, as some of the studies
were carried out for less than 2 weeks.

Negative mineral balances as a result of oxalic acid and
fiber together in the diet may not necessarily be a problem, as it
is not too likely that spinach would be included in even a high
fiber diet as often as 3 or 4 times a week. However, it is
possible that the frequent inclusion of foods containing oxalic
acid in the diet might be of concern where diets are high in fiber
and low in mineral content. Singh et al. (24) reported that the
normal rural diet in the Udaipur region of India is low in oxalic
acid but that its intake rises sharply in the season when amaranth, purslane, pigweed, and spinach are abundant. In that
region, oxalic acid intake was consistently higher for the upper
income group than for the low income group. Tabekhia et al. (25)
found a high level of oxalate in relation to calcium in Jew's
mallow and purslane, which are two important leafy vegetables in
the Egyptian diet.

Further study of the effects of different levels of oxalic
acid and fiber intake on mineral balances is recommended.

Literature Cited

1. Reinhold, J. G.; Ismail-Beigi, F.; Faradji, B.   Nutr.
    Rep. Int. 1975, 12, 75-85.
2. Ismail-Beigi, F.; Faradji, B.; Reinhold, J. G.   Am. J.
    Clin. Nutr. 1977, 30, 1721-5.
3. Camire, A. L.; Clydesdale, F. M.   J. Food Sci. 1981, 46,
    548-51.
4. Kelsay, J. L.   Cereal Chem. 1981, 58, 2-5.
5. Hodgkinson, A.; Zarembski, P. M.   Calc. Tiss. Res. 1968, 2,
    115-32.
6. Weast, R. C.; Astle, M. J.   "CRC Handbook of Chemistry and
    Physics"; 61 ed.; CRC Press, Inc.: Boca Raton, FL, 1980-81.

7. Kohman, E. P. J. Nutr. 1939, 18, 233-46.
8. Majumdar, B. N.; De, N. K. Ind. J. Med. Res. 1938, 25, 671-5.
9. Andrews, J. C.; Viser, E. T. Food Res. 1951, 16, 306-12.
10. Zarembski, P. M.; Hodgkinson, A. Brit. J. Nutr. 1962, 16, 627-34.
11. Kaul, S.; Verma, S. L. Ind. J. Med. Res. 1967, 55, 274-8.
12. Gontzea, I.; Sutzescu, P. "Natural antinutritive substances in foodstuffs and forages"; S. Karger: Basel, Switzerland, 1968.
13. Walker, A. R. P.; Walker, B. F.; Wadvalla, M. Ecol. Fd. Nutr. 1975, 4, 125-30.
14. Pingle, U.; Ramasastri, B. V. Brit. J. Nutr. 1978, 39, 119-25.
15. Pingle, U.; Ramasastri, B. V. Brit. J. Nutr. 1978, 40, 591-4.
16. Welch, R. M.; House, W. A.; Van Campen, D. J. Nutr. 1977, 107, 929-33.
17. Kelsay, J. L.; Behall, K. M.; Prather, E. S. Am. J. Clin. Nutr. 1978, 31, 1149-53.
18. Kelsay, J. L.; Goering, H. K.; Behall, K. M.; Prather, E. S. Am. J. Clin. Nutr. 1981, 34, 1949-52.
19. Kelsay, J. L.; Jacob, R. A.; Prather, E. S. Am. J. Clin. Nutr. 1979, 32, 2307-11.
20. Kelsay, J. L.; Clark, W. M.; Herbst, B. J.; Prather, E. S. J. Agr. Food Chem. 1981, 29, 461-5.
21. Kelsay, J. L.; Prather, E. S. Fed. Proc. 1981, 40, 854.
22. Kojima, N.; Wallace, D.; Bates, G. W. Am. J. Clin. Nutr. 1981, 34, 1392-1401.
23. Van Campen, D. R.; Welch, R. M. J. Nutr. 1980, 110, 1618-21.
24. Singh, P. P.; Kothari, L. K.; Sherma, D. C.; Saxena, S. N. Am. J. Clin. Nutr. 1972, 25, 1147-52.
25. Tabekhia, M. M.; Toma, R. B.; El-Mahdy, A. R. Nutr. Rep. Int. 1978, 18, 611-6.

RECEIVED October 13, 1982

The Role of Phytate in Zinc Bioavailability and Homeostasis

D. OBERLEAS

University of Kentucky, Department of Nutrition and Food Science, Lexington, KY 40506

Phytate, myo inositol hexakis (dihydrogen phosphate) is found in all plant seeds, many roots and tubers. Its synthesis follows pollination and its content increases with maturity. The greatest concentration is found in legume seeds, bran and germ of cereal grains. Small amounts are found in many fruits and vegetables except stem and leafy vegetables. Phytate has been shown to be associated with protein, as magnesium salts in sesame, and is water soluble in corn germ. Phytate complexes with divalent elements with varying degrees of tenacity. The solubility of these complexes varies with pH. Zinc phytate is least soluble at pH 6 and is less soluble than calcium or other mineral complexes at this pH. Kinetic synergism of calcium and zinc with phytate causes a complexation less soluble than either separately. Saliva and pancreatic fluid secrete large quantities of zinc equivalent to as much as three times the dietary intake which is also vulnerable to phytate complexation. The mechanism of phytate action in the gastrointestinal tract is related to complexation and subsequent prevention of absorption and reabsorption of zinc. The complexation can be equated to a phytate:zinc molar ratio and the relative hazard may be subsequently estimated from such data.

A compound later to be identified as myo-inositol hexakis (dihydrogen phosphate) was first isolated from plant seeds by Pfeffer in 1872 (1) (Figure 1). The relative concentration and widespread distribution in plant seeds was first described by Palladin (2) in 1895 and identity as an inositol compound in 1897 by Winterstein (3). Its identity as a hexaphosphate ester of inositol was confirmed early in this century (4) and synthesis was accomplished in 1919 by Posternak (5). The biosynthesis of

0097-6156/83/0210-0145$06.00/0

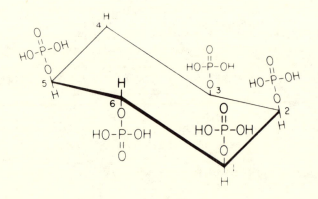

Figure 1. *1,2,3,4,5,6-Hexakis(phosphonooxo)cyclohexane [phytate] (39).*

phytate has been studied in corn (6, 7) and peas (8). Pollination was shown to supply the essential zymogen activator necessary for phytate formation and the concentration increased to maturity.

Phytate exists as different complexes in different seeds. In corn, phytate is contained primarily in the germ in a water soluble form (9). Though it seems unlikely that such a strong acid would be free, complexation with potassium, for example, would render water solubility without being identifiable by current methodology. Also complexation with proteins' which were either water soluble or whose isoelectric point were above or below the pH of water would render solubility. In legumes, phytate has been shown to be associated with protein (9, 10). This association is greatest at the isoelectric point of the protein and being readily dissociated at pH's above or below the isoelectric point (10, 11). Phytate association in cereal grains is less well defined but is contained in significant concentrations both in the bran and germ (12). There appears to be at least one ferric ion associated in the otherwise soluble phytate complex in wheat bran (13). Phytate in sesame seed appears to be the most unique and least soluble of all seeds. In this case magnesium appears to be the predominant cation (9).

The essentiality of zinc was first demonstrated in 1869 by Raulin (14) using Aspergillus niger. Attempts by Bertrand and Benzon in 1922 (15) to deprive mice of zinc were inconclusive and not until 1934 (16, 17) was zinc considered essential for animals. These early studies incorporated highly purified diets containing less than 2 mg zinc per kg diet. The practical significance of zinc was not realized until 1955 when Tucker and Salmon (18) demonstrated that supplemental zinc would prevent or cure porcine parakeratosis which was prevalent at that time. Shortly thereafter O'Dell and Savage (19) noted that zinc was less available to the chick from plant proteins than animal proteins. This was quickly confirmed by several investigators (20 - 23).

Phytate, associated with legume seed proteins, is one basic difference between plant and animal proteins. The addition of phytate to a casein-gelatin diet for chicks was shown to produce the same symptoms of zinc deficiency as did a soybean protein diet (24). This is illustrated in Figure 2. This work was extended to show a calcium, phytate, zinc interrelationship (25). Supporting observations have subsequently been made in many species, including swine (26), dogs (27), Japanese quail (28), rats (29), and rainbow trout (30) though frequently the observers were not aware of the role of phytate in developing the zinc deficiency syndrome. A typical growth curve for rats is shown in Figure 3.

In the first report of clinical zinc deficiency in humans (31, 32) the significant dietary consideration, not fully appreciated at that time, was that the village population subsisted primarily on unleavened whole wheat bread or bread and beans and very little animal protein was consumed by this population.

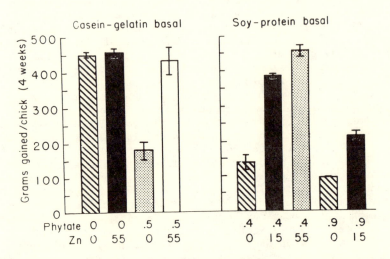

Figure 2. Chick growth at 4 weeks as affected by phytate and supplemental zinc. Basal diets contained 9 mg/kg zinc. (Reproduced with permission from Ref. 40.)

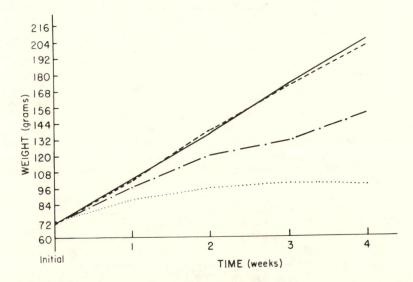

Figure 3. Growth curves of rats fed casein-based diets containing phytate and calcium. Each line represents the mean of 11 animals. Basal diets contained 6 mg/kg zinc. Key: ---, 0.8% Ca, 0% phytate; ——, 1.6% Ca, 0% phytate; — · —, 0.8% Ca, 1% phytate; · · ·, 1.6% Ca, 1% phytate. (Reproduced with permission from Ref. 40.)

Geophagia was also reported for the subjects in Iran and an analysis of a sample of clay eaten by some of the subjects indicated that the clay contained a very high concentration of calcium.

Chemical Relationships

The interest in the chemical interrelationships of phytate initially centered around its complexation with calcium and subsequent effect on the availability of calcium (33). This led to the classical studies by Hoff-Jorgenson (34). He determined eight of the twelve dissociation constants of phytic acid and using these data calculated the solubility products of penta-calcium phytate as between 10^{-19} and 10^{-23}. A decade later with the use of neutralization curves and conductivity measurements it was concluded that phytic acid contained 6 strong acid groups which were completely dissociated in solution (pk = 1.84), 2 weak acid functions (pk = 6.3) and 4 very weak acid protons (pk = 9.7) (34).

Phytate forms salts/complexes with most heavy metals. The relative solubility of the complex is dependent on pH as well as on the presence of a secondary cation, of which only calcium has been studied (29, 35). Of greatest interest are the cations which are complexed most tightly at the physiological range of pH's. Complexes of single divalent cations, at pH 7.4, formed in the following decreasing order: Cu^{++} > Zn^{++} > Co^{++} > Mn^{++} > Fe^{++} > Ca^{++} (36, 37). It is interesting that calcium, which has until recent years received the most attention, is at the lower end of this series. Ferric phytate is known to be least soluble in dilute acid. However as a tertiary component in the complexation of zinc (26, 29) and other divalent elements (35), calcium has been shown to promote a synergism that is unique in chemical kinetics. This was first studied in an in vitro model while searching for a mechanism to explain the experimental observations of calcium and/or phytate as causative agents in zinc deficiency (38).

Figure 4 illustrates the quantity of precipitate formed when equimolar concentrations of phytate, zinc and/or calcium (1:1 or 2:1) are mixed in an open vessel and pH's adjusted first to less than 3, than carefully to the appropriate pH. The pH range between 3 and 9 was selected to encompass the physiologically important range. The results indicate that 1) calcium and phytate, in equimolar concentration, are quite soluble at all pH's under these conditions; 2) zinc phytate is less soluble than calcium phytate and at pH 6 is less soluble than calcium at twice the molar concentration; 3) zinc, calcium and phytate in all combinations tested was less soluble than either zinc or calcium phytate or the sum of these alone at pH 6. The pH 6 is very important physiologically because this is the approximate pH of the duodenum and upper jejunum, an area of the gastrointestinal tract in which zinc must be absorbed. At pH 6 and a 2:1:1 calcium:zinc:phytate molar ratio, 98% of the zinc was in the precipitate (28, 38, 39).

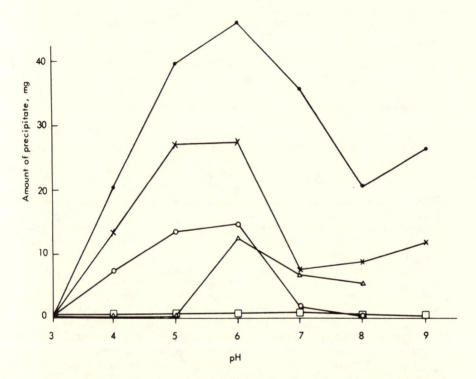

Figure 4.   Quantity of precipitate formed at varying molar ratios and pH's. Key:
□, *Ca:phytate (1:1);* △, *Ca:phytate (2:1);* ○, *Zn:phytate (1:1);* ×, *Ca:Zn:phytate*
(1:1:1); ●, *Ca:Zn:phytate (2:1:1). (Reproduced with permission from Ref. 40.)*

Zinc is a trace element and both calcium and phytate are present in many foods in macro quantities; therefore, an in vitro model was developed to study the quantity of zinc available for absorption with these more physiological ratios. The results, Figure 5, indicate that as the ratio of calcium:phytate increases there is a decrease in the uncomplexed zinc in solution which would be available for absorption (28, 38, 40). EDTA added into this same model increased the soluble zinc (38) indicating that soluble and absorbable chelating compounds may compete with phytate and make some zinc available for absorption or reabsorption. The implications of the studies described above are that the interaction of phytate with zinc and calcium involves a chemical rather than a physiological reaction.

With the large differential between the molecular weight of phytate (660) and the atomic weight of zinc (65.4), weight or percentage comparisons would be of little value. Phytate:zinc molar ratio reduces both of these components to a common denominator which represents a meaningful, chemical comparison. It is apparent that some means was needed to express this relationship on a chemical basis that also has physiological implications. The use of phytate:zinc molar ratio was tested on data from several experiments in which the dietary model was indentical except for the concentrations of phytate and zinc. Since growth rate depression is an early, sensitive and easily measured parameter, it was quite suitable for the dependent variable. The data, Table I, shows the inverse relationship between decreasing phytate:zinc molar ratio and average growth rate of these rats within 4 weeks. The severity of clinical symptoms are directly

Table I.   VARIABILITY OF ZINC DEFICIENCY

| Diet | Dietary Varient | | Phytate/Zinc | Gain/Wk $\pm$ SD |
	Phytate (%)	Zinc (mg/kg)	(molar)	(4 Weeks)
1	1.0	2*	495.5	8.8 $\pm$ 3.0 (24)
2	1.0	15	66.1	12.5 $\pm$ 4.1 (27)
3	0.4	15	26.1	15.3 $\pm$ 4.6 (18)
4	0.4	70	5.7	37.5 $\pm$ 7.0 (47)
5	0.4	125	3.2	48.5 $\pm$ 4.2 (30)

*Protein EDTA washed.

From Oberleas, D. (39).

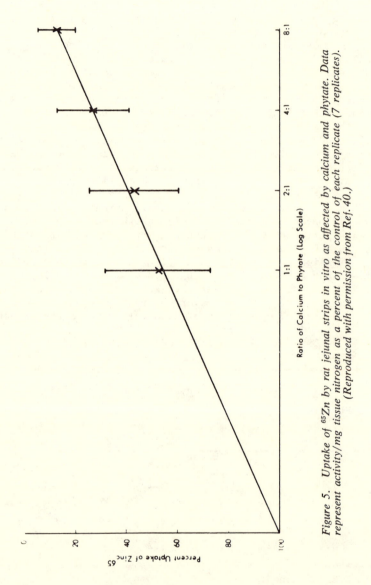

Figure 5. Uptake of ^{65}Zn by rat jejunal strips in vitro as affected by calcium and phytate. Data represent activity/mg tissue nitrogen as a percent of the control of each replicate (7 replicates). (Reproduced with permission from Ref. 40.)

related to the phytate:zinc molar ratio (39). The diets utilized in these experiments contained 1.6 percent calcium which accentuates the experimental effect of phytate but is considerably higher than the typical human intake. Three laboratories have subsequently and independently confirmed the validity of the phytate:zinc molar ratio utilizing dietary models with more reasonable calcium levels (41, 42, 43). All three laboratories have concluded that with moderate calcium intake, molar ratios less than 10 are likely to provide adequate available zinc and molar ratios greater than 10 are associated with symptoms of zinc deficiency; namely, growth rate depression. Though the determination of this critical molar ratio was made in rats, it represents an expression of a chemical relationship and thus applies to all monogastric species for similar dietary phytate:-zinc molar ratios. The molar ratio of 10 also compares favorably with the need for a molar ratio between 3 and 6 at the higher levels of calcium intake to provide adequate available zinc.

Homeostatic Fluxes of Zinc

Frequently ignored in many studies are the relatively large quantities of zinc which are recycled into the gastrointestinal tract. The most widely studied are the relatively large secretions of zinc in the pancreatic fluid (44, 45, 46). The pancreatic contribution to the duodenum may be more than three (3) times the dietary intake (44, 45). Some contribution is also made thru the bile (44, 46). Zinc is also secreted in sizeable quantities in the saliva (47). With such a large contribution of zinc to the total gastrointestinal pool, much of the secreted zinc must be reabsorbed to prevent the body from experiencing a perpetual zinc deficit.

The relative vulnerability of the secreted zinc to phytate complexation has only recently been demonstrated. The injection of zinc deficient rats intraperitoneally with a tracer dose of ^{65}zinc allows a portion of this zinc to be in equilibrium with the endogenous metabolic pool. This zinc then is secreted thru the saliva, pancreatic fluid, and bile. Those animals maintained on the phytate containing soy protein contained 2-4 times the radioactivity of the animals fed a casein protein diet (Table II). Therefore, not only does phytate affect the bioavailability of dietary zinc but also the reabsorption of endogenous zinc and thus has a net effect on zinc homeostasis. Since this total phytate effect cannot be measured by labeling only the dietary pool, the expression of the net effect as the phytate:zinc molar ratio is the most sensitive and accurate method of estimating the relative risk of zinc deficiency in any individual or population.

TABLE II. RATIO OF ^{65}ZINC EXCRETED BY RATS ON SOY PROTEIN
DIET VS CASEIN PROTEIN DIET FOLLOWING INTRAPERITONEAL INJECTION

Days	1	3	5	7	9	11	13
1.6 % Ca	1.5	1.4	1.8	3.9	3.3	2.8	2.8
0.8 % Ca	1.3	1.6	1.6	2.3	3.3	2.5	2.5

Mechanism of Action

In order to formulate a mechanism of action of phytate on
zinc homeostasis, certain conditions must be fulfilled: 1) The
process must occur in the gastrointestinal tract since phytate is
not absorbed except for small amounts by birds; 2) calcium must
be a tertiary component in the total process but there must be
some reaction without excessive amounts of calcium; 3) certain
chelating compounds such as EDTA must be capable of competing
with the process and make some zinc available for absorption or
reabsorption and 4) There must be some explanation for the data
which indicates that 40% or more of the dietary pool may be
available for absorption (48, 49). All these conditions are
satisfied by the following formula which is an expression of the
"Law of Mass Action":

$$Zn^{++} + Phytate \quad \dfrac{Ca^{++}}{EDTA} \quad Zn(Ca)Phytate$$

One must recognize that the total zinc pool available for satis-
fying the conditions of this reaction must include the endogenous
secreted zinc which may represent 3-4 times the dietary intake
even when an adequate level of zinc is consumed. Whereas phytate
is a polyfunctional compound and the solubility product of such a
mixed salt is difficult to calculate, the concept of expressing
the phytate and zinc as molar ratios is simple to calculate and
useful in the determination of relative zinc status (50).

Potential Population Hazards

The earliest tests of the phytate:zinc molar ratio concept
in humans was provided by data supplied by Reinhold (51, 52). He
analyzed a large number of samples of Middle Eastern bread for
phytate and zinc. These breads were of two types; "Bazari and
Sangak" which are leavened breads sold in the urban areas and
"Tanok" which is an unleavened bread consumed in the villages of
the Middle East (Table III). For many of the villagers, "Tanok",

constitutes over 90% of their diet. It is from these village populations that clinical zinc deficiency was first described (31) and is easily diagnosed (53). At a zinc concentration of 30 mg/kg (Reinhold, personal communication) the unleavened "Tanok" has a phytate:zinc molar ratio of about 23 whereas the leavened bread had molar ratios of about 10-12. The latter would imply an element of risk but not to the extent of exhibiting clinical symptoms. Since the "tanok" bread constitutes the major portion of the village diets, if an estimate of consumption of bread were 1 kg/day, clinical zinc deficiency was prevalent at a molar ratio of 23 in spite of a zinc intake of 30 mg/day. The latter would be twice the current Recommended Dietary Allowance (RDA) in the United States.

TABLE III.   PHYTATE CONTENT OF MIDDLE EASTERN BREADS

	Bazari	Sangak	Tanok
Samples (N)	56	50	126
Phytate (mg/100 g)	326	388	684
Phytate (m moles)	4.94	5.88	10.36
Zinc	Estimated at 30 mg/kg for all breads (0.46 m moles)		
Phytate/Zinc (molar)	10.8	12.8	22.6

From: Oberleas (39).

A current study of dietary recalls from EFNEP subjects on self-selected diets from three areas of Kentucky indicate that from a systematically selected sample of 1294 subjects approximately 800 had phytate:zinc molar ratios greater than 10 and about 600 had ratios greater than 15. A small group of subjects had molar ratios greater than 15 and zinc intakes greater than 15 mg/day.

Beyond this, one has but to look at typical diets consumed throughout the world and imagine that zinc deficiency, although not always expressing itself with overt clinical symptoms, may affect a very large segment of the world's population. Particularily vulnerable are the poorer segments of the population which must depend on cereal grains and legume seeds as the major source of protein and calories. Within the United States, the most "at risk" population may be represented by adolescents who frequently include cereals, peanut butter and pasta as major dietary components, "vegans", or those substituting, soybean based synthetic meats for animal protein.

As analytical techniques for estimating phytate become available, computer data bases can be updated to contain accurate estimates of the phytate and zinc concentrations of foods. When good data bases become established a simple and noninvasive technique of calculating phytate:zinc molar ratio from dietary recalls will provide a sensitive and accurate estimate of the relative risk of zinc deficiency for any population. Since this represents a physical constant the molar ratio of 10 or less represents and adequate standard provided some minimal intake of approximately 5 mg of zinc/day is consumed. Likewise, the RDA for zinc, currently at a somewhat arbitrary 15 mg/day needs to be modified to reflect the effect of phytate on zinc homeostasis as expressed by the phytate:zinc molar ratio. Zinc from almost any mineral source is readily available for absorption (54). Therefore, in the absence of phytate, zinc deficiency represents little more than an academic curiosity; the effect exhibited by phytate makes zinc deficiency a real and prevalent problem.

Acknowledgment

This was supported in part by the Science and Education Administration of the U.S. Department of Agriculture under Grant No. 5901-0410-8-0076-0 from the Competive Research Grants Office.

Literature Cited

1. Pfeffer, E. Pringsheims Jb. Wiss. Bot. 1872, 8, 429,475.
2. Palladin, W. Z. Biol. 1895, 31, 199.
3. Winterstein, E. Ber. dtsch. chem. Ges. 1897, II, 2299.
4. Suzuki, U.; Yoshimura, K.; Takaishi, M. Coll. Agric. Bull., Tokyo Imp. Univ. 1907, 7, 503.
5. Posternak, S. Compt. Rend. 1919, 169, 138-40.
6. De Turk, E.E.; Holbert, J.R.; Hawk, B.W. J. Agric. Res. 1933, 46, 121-41.
7. Earley, E.B.; De Turk, E.E. J. Am. Soc. Agron. 1944, 36, 803-14.
8. Fowler, H.D. J. Sci. Food Agric. 1957, 8, 333-41.
9. O'Dell, B.L.; deBoland, A.R.; Koirtyohann, S.H. J. Agric. Food Chem. 1972, 20, 718-21.
10. Fontaine, T.D.; Pons, W.A., Jr.; Irving, G.W., Jr. 1946, 164, 487-507.
11. Smith, A.K.; Rackis, J.J. J. Am. Chem. Soc. 1957, 79, 633-7.
12. deBoland, A.R.; Garner, G.B.; O'Dell, B.L. J. Agric. Food Chem. 1975, 23, 1186-9.
13. Morris, E.R.; Ellis, R. J. Nutr. 1976, 106, 753-60.
14. Raulin, J. Ann. Sci. Nat. Bot. Vegetale 1869, 11, 93.
15. Bertrand, G.; Benzon B. Compt. Rend. 1922, 175, 289-92.
16. Todd, W.R.; Elvehjem, C.A.; Hart E.B. Am. J. Physiol. 1934, 107, 146-56.
17. Stirn, F.E.; Elvehjem, C.A.; Hart E.B. J. Biol. Chem. 1935, 109, 347-59.
18. Tucker, H.F.; Salmon, W.D. Proc. Soc. Exp. Biol. Med. 1955, 88, 613-6.

19. O'Dell, B.L.; Savage, J.E. Poul. Sci. 1957, 36, 459-60.
20. Morrison, A.B.; Sarett, H.P. J. Nutr. 1958, 65, 267-80.
21. Kratzer, F.H.; Allred, J.B.; Davis, P.N.; Marshall, B.J.; Vohra, P.N. J. Nutr. 1959, 68, 313-22.
22. Ziegler, T.R.; Leach, R.M., Jr.; Norris, L.C.; Scott, M.L. Poul. Sci. 1961, 40, 1584-93.
23. Smith, W.H.; Plumlee, M.P.; Beeson, W.M. J. Anim. Sci. 1692, 21, 399-405.
24. O'Dell, B.L.; Savage, J.E. Proc. Soc. Exp. Biol. Med. 1960, 103, 304-6.
25. O'Dell, B.L.; Yohe, J.M.; Savage, J.E. Poul. Sci. 1964, 43, 415-19.
26. Oberleas, D.; Muhrer, M.E.; O'Dell, B.L. J. Anim. Sci. 1962, 21, 57-61.
27. Robertson, B.T.; Burns, M.J. Am. J. Vet. Res. 1963, 24, 997-1002.
28. Fox, M.R.S.; Harrison, B.N. Proc. Soc. Exp. Biol. Med. 1964, 116, 256-61.
29. Oberleas, D.; Muhrer, M.E.; O'Dell, B.L. J. Nutr. 1966, 90, 56-62.
30. Ketola, H.G. J. Nutr. 1979, 109, 965-9.

31. Prasad, A.S.; Halsted, J.A.; Nadimi, M. Am. J. Med. 1961, 31, 532-46.
32. Prasad, A.S.; Miale, A., Jr;, Farid, Z.; Sandstead, H.H.; Schulert, A.R.; Darby, W.J. AMA Arch. Int. Med. 1963, 111, 407-28.
33. McCance, R.A.; Widdowson, E.M. Biochem. J. 1935, 29, 2694-99.
34. Barre', R.; Courtois, J.E.; Wormser, G. Bull. Soc. Chim. Biol. 1954, 36, 455-74.
35. Oberleas, D.; Moody, N. Trace Element Metabolism in Man and Animals-IV. (J. McC. Howell, J.M. Gawthorne, and C.L. White, eds.), Australian Academy of Science, Canberra, ACT., 1981, pp. 129-132.
36. Maddariah, V.T.; Kurnick, A.A.; Reid, B.L. Proc. Soc. Exp. Biol. Med. 1964, 115, 391-3.
37. Vohra, P.; Gray, G.A.; Kratzer, F.H. Proc. Soc. Exp. Biol. Med. 1965, 120, 447-9.
38. Oberleas, D. Ph.D. Dissertation, U. of Missouri, 1964, pp. 143-54.
39. Oberleas, D. Proc. Western Hemisphere Nutr. Congress-IV (P.L. White and N. Selvey, eds.) Publishing Sciences Group, Inc., Acton, MA, 1975, pp. 156-61.
40. Oberleas, D.; Muhrer, M.E.; O'Dell, B.L. Zinc Metabolism (A.S. Prasad, ed.) C.C. Thomas, Springfield, IL. 1966, pp. 225-38.
41. Davies, N.T.; Olpin, S.E. Br. J. Nutr. 1979, 41, 591-603.
42. Morris, E.; Ellis, R. J. Nutr. 1980, 110, 1037-45.
43. Lo, G.S.; Settle, S.L.; Steinke, F.H.; Hopkins, D.T. J. Nutr. 1981, 111, 2223-35.

44. Pekas, J.C. Am. J. Physiol. 1966, 211, 407-13.
45. Miller, J.K.; Cragle, R.G. J. Dairy Sci. 1965, 48, 370-3.
46. Sullivan, J.F.; Williams, R.V.; Wisecarver, J.; Etzel, K.; Jetton, M.M.; Magee, D.F. Proc. Soc. Exp. Biol. Med. 1981, 166, 39-43.
47. Freeland-Graves, J.H.; Hendrickson, P.J.; Ebangit, M.L.; Snowden, J.V. Am. J. Clin. Nutr. 1981, 34, 312-321.
48. O'Dell, B.L.; Burpo, C.E.; Savage, J.E. J. Nutr. 1972, 102, 653-60.
49. Evans, G.W.; Johnson, P.E. Am. J. Clin. Nutr. 1977, 30, 873-78.
50. Oberleas, D.; Harland, B.F. J. Am. Diet. Assoc. 1981, 79, 433-6.
51. Reinhold, J.G. Am. J. Clin. Nutr. 1971, 24, 1204-06.
52. Reinhold, J.G. Ecol. Food Nutr. 1972, 1, 187-82.
53. Ronaghy, H.A. Pahlavi Med. J. (Shiraz) 1970, 1, 29.
54. Edwards, H.M., Jr. J. Nutr. 1959, 69, 306-8.

RECEIVED October 13, 1982

Dietary Phytate/Zinc Molar Ratio and Zinc Balance in Humans

EUGENE R. MORRIS and REX ELLIS

U.S. Department of Agriculture, Agricultural Research Service, Beltsville
Human Nutrition Research Center, Beltsville Agricultural Research Center,
Beltsville, MD 20705

The effect on zinc balance of a 10-fold difference
in dietary molar ratio of phytate/zinc was tested in
a metabolic balance study with 10 adult men. The
mean zinc balance was 2.74 mg per day when the
dietary molar ratio of phytate/zinc was about 12 and
2.0 mg per day when the ratio was about 1. Menus
consisted of foods commonly consumed in the United
States. The mean daily intake of zinc was 17 mg and
of neutral detergent fiber 16 g. The molar ratios
of phytate/zinc were attained by using 36 g per day
of whole or dephytinized wheat bran. Analysis of
hospital and self-chosen diets indicate that the
majority of the United States population consume
diets with molar ratios of phytate/zinc less than
10, but which provide less than the recommended
dietary allowance of zinc. The balance results are
discussed in relation to magnitude of the zinc
intake, the type of food consumed and the role of
adaptive responses in maintaining adequate zinc
nutriture.

Zinc is a required mineral nutrient in the diet of animals
and the signs and effects of a dietary deficiency have been
described in several species (1). Nutritional zinc deficiency
has also been described in humans (2). Signs of nutritional zinc
deficiency may be manifest even though the individual is consuming
an amount of dietary zinc that exceeds the usually designated
requirement (3). Thus, bioavailability of dietary zinc is a
factor that must be evaluated in considering adequacy of dietary
intake (4). Phytic acid and dietary fiber in foods have been implicated
as important determinants of dietary zinc bioavailability (5,6,7).
For rats, the molar ratio of phytic acid to zinc is a good
predictor of bioavailability of dietary zinc when semipurified
type diets are fed in which $ZnSO_4$ and either sodium phytate or

soy protein are the dietary source of zinc and phytic acid, respectively (8,9,10). Little information is available to ascertain the influence of the dietary molar ratio of phytate/ zinc on zinc nutrition of humans. Dietary fiber from wheat bran in the diet of rats does not seem to influence bioavailability of zinc but the phytate does (11,12,13). Differentiation between the effect of phytate and fiber is difficult in bioavailability studies with wheat bran or other cereal and legume products that contain both phytate and dietary fiber. By the action of the endogenous phytase, wheat bran can be dephytinized without changing the dietary fiber concentration (13). Although there may be an as yet unidentified chemical change in the dietary fibèr, enzymatically dephytinized wheat bran provides a relatively high fiber food product from cereal grain that is low in phytate.

We conducted a human metabolic balance study to test the concept of dietary phytate/zinc molar ratio as a predictor of zinc bioavailability to humans. Using unaltered or enzymatically dephytinized wheat bran with ordinary foods we attained phytate zinc molar ratios of about 1 and 12 with relatively high intakes of dietary fiber in the menus and found no difference in zinc balance. Retrospectively, the result may be qualified on the basis of the magnitude of the zinc intake and possible adaptive or homestatic responses over the period of the study. A second study was then conducted and a wider range of phytate/zinc molar ratio was provided than in the first study. We will briefly outline the first study, give a progress report on the second study and, with some information on phytate intakes obtained by our laboratory, discuss the nutritional implication.

Balance Study I

Study Protocol and Materials. Ten healthy men ranging in age from 23 to 48 years volunteered as subjects for the study. Each gave informed consent and was given a physical examination prior to the study. An outline of the study protocol is shown in Figure 1. Each phytate/zinc molar ratio was consumed for 15 days, three repeats of the 5-day menu cycle. Stool and urine collections were pooled for the periods as indicated in Figure 1. A fasting blood sample for serum zinc analysis was taken on days 1, 16 and 31.

The menus consisted of foods routinely consumed in the United States. An example of one day's menu is given in Table I. Different caloric intakes were provided, to maintain each subjects weight, by varying the amount of the non-muffin items in increments of 10% using the 3200 kcal meals as base. Each subject consumed two bran muffins with each meal, 6 g of bran per muffin. All food was supplied and cooked by the Human Study Facility of the Beltsville Human Nutrition Research Center. Monday through Friday meals were consumed at the Human Study Facility and take-home packs were supplied for weekends and one holiday.

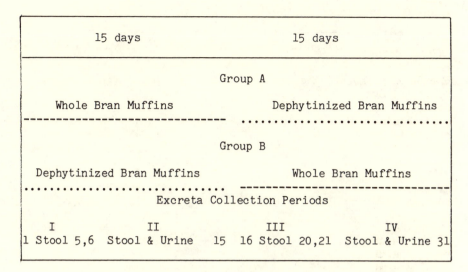

Figure 1.   Outline of experimental protocol. Menus repeated in 5-day cycles. Two bran muffins consumed with each meal (6 g of bran/muffin). Five subjects each subgroup, A & B. Brilliant blue dye (50 mg) was given with breakfast on days 1, 6, 16, 21, and 31 (lower line) to aid in demarcation of stool.

Deionized water, instant tea and coffee were supplied for drinking, but no other beverages were allowed.

The wheat bran was milled by the USDA Wheat Quality Laboratory, Fargo, ND from a single lot of hard red spring wheat. In our laboratory at Beltsville, one-half of the bran was refrigerated until used (whole bran) and one-half was dephytinized by endogenous phytase activity (12). The bran was suspended in deionized water (1 g bran to 10 ml water) and incubated at 37° with constant shaking overnight or about 16 hours. After incubation the entire water suspension of dephytinized bran was freeze-dried, retaining the mineral nutrients and hydrolysis products, and stored refrigerated until use. The brans were baked into muffins, 6 g of bran per muffin, and each subject consumed 6 muffins or 36 g of bran each day. Each subject consumed each type of bran muffin over the course of the study, but 5 (group A) consumed whole bran muffins and 5 (group B) consumed dephytinized bran muffins for the first 15 days and the alternate type of bran muffin the second 15 days.

Daily composites were made of the 3200 kcal menu items (muffins were composited separately), homogenized and freeze-dried for analysis. Stools were freeze-dried and throughly mixed before sampling for analysis. A 50 ml aliquot of urine was dried in a porcelain crucible and dry ashed. Zinc analysis was by flame

Table I. Example of One Day's Menu[1]

BREAKFAST

Orange/Grapefruit Juice	124 grams
Scrambled Egg	50
Margarine	5
Sausage Links	42
Milk, 2%	246
Muffins	120
Jelly[2]	66
Margarine[2]	40
Sugar[2]	10

LUNCH

Turkey Roast	90
Macaroni	70
Mayonnaise	15
Cranberry Juice	126
Ice Cream	100
Muffins	120

DINNER

Steak	120
Boiled Potato	50
Stewed Tomato	50
Pumpkin Pie	100
Milk, 2%	246
Grape Juice	253
Muffins	120

---

[1]3200 kcal, 132 g protein, 40% calories from fat, P/S - 0.66.
[2]Given with breakfast and used throughout the day.

atomic absorption spectroscopy of 1 N HCl solution of dry ashed
diet or stool composite. Phytate analysis was by the method of
Ellis et al. (14). Neutral detergent fiber was determined by the
approved method of the American Association of Cereal Chemists
(15).

Results

     Results pertaining to zinc will be summarized here.
Information concerning other mineral nutrients have been
published in part in abstracts and a short communication
(16,17,18).

Nutrient intake. Table II summarizes the daily intakes of
zinc and phytic acid. The range of zinc intake is the result of
the different caloric levels consumed. About one-third of the
zinc consumed was contained in the muffins. About 95% of the
phytic acid was derived from the muffins and, since the same
number of muffins was eaten regardless of caloric level, there
was no difference due to the caloric level in phytic acid intakes.
There was, however, a ten-fold difference in phytic acid intakes
by type of bran muffin. The mean calculated molar ratio of
phytate/zinc was about 12 when whole bran muffins were consumed
and just greater than 1 when dephytinized bran muffins were

Table II. Daily Intakes of Zinc and Phytic Acid

Muffin Type	Zinc	Phytic Acid	Phy/Zn Molar Ratio
	mg	g	
Whole Bran	16.8 (14.7 - 18.9)[1]	2.0	11.8 (10.5 - 13.6)[1]
Dephytinized bran	17.1	0.2	1.2

[1]Range

consumed. An equal amount of neutral detergent fiber was
supplied by each type of bran muffin. By analysis, a mean of 16
g of neutral detergent fiber was consumed daily.
    Zinc balance. The zinc balances for the last 10 days of
each 15 day period (periods II and IV, Figure 1) are shown in
Figure 2. All subjects were in positive zinc balance regardless
of the type of bran muffin consumed. Of the 10 individual
subjects, the zinc balance for 8 was greatest when consuming the
whole bran muffins. Only two individuals, one in each subgroup,
were in more positive zinc balance when consuming dephytinized
bran muffins. Overall the mean balance was 0.7 mg per day
greater when eating whole bran muffins than when eating dephytinized bran muffins. This difference was not statistically
significant.
    Values for intake minus fecal excretion of zinc are presented
in Table III. The values are mg/day for the 5 days of periods I
and III and the 10 days of periods II and IV. Regardless of the
sequence in which the two types of bran were consumed, values
for intake minus fecal exretion were greater for the last 10
days each type of muffin was consumed than for the first 5 days,

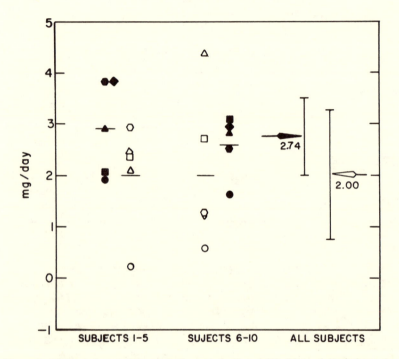

Figure 2. Zinc balances for metabolic periods II and IV. Mean daily balance for individual subjects shown in the sequence of type of bran muffin consumed. Within each subgroup, each individual is represented by a different shaped figure. Means are shown by horizontal lines for each subgroup and the arrows for all subjects. Key: solid symbols, whole bran; and open symbols, dephytinized bran.

periods II and IV compared to I and III, respectively. The
magnitude of the S.D. for periods I and III verify that some
individuals of each subgroup excreted more zinc in the stool
than was consumed, especially subgroup A during period I. The
first 5 days whole bran muffins were consumed the intake minus
fecal excretion values were always less in magnitude than the
first 5 days dephytinized bran muffins were consumed, group A
periods I vs. III and B periods III vs. I.

Table III. Intake Minus Fecal Excretion of Zinc

Subject Subgroup	Excreta Collection Period[2]			
	I	II	III	IV
			mg/day[1]		
A	(+)[3]		(-)		
	-1.4+1.8	3.2+0.9	1.8+3.6	2.4+1.1
B	(-)		(+)		
	1.2+1.3	2.4+1.6	0.9+2.0	3.0+0.4

[1]Mean + S.D.
[2]See Figure 1.
[3](+) Whole bran muffins, (-) dephytinized bran muffins.

Serum zinc response. Table IV is a summary of the initial
serum zinc values and at the end of 15 days consuming each type of
bran muffins. The tendency was toward higher serum zinc values
during the study. The initial mean for subgroup B was lower than
A's and the increases during the study were statistically
significant. Averaging both subgroups, the mean serum zinc was
10 μg/100 ml greater after consuming the dephytinized bran for 15
days, but the means were not significantly different.
Discussion. Oberleas discusses in these proceedings some of
the factors that influence the solubility of zinc in the presence
of phytate. Rats can utilize the zinc of water insoluble zinc
phytate or calcium-zinc phytate complexes in the absence of any
added phytate (10,19). Growth, but not femur zinc response was
equivalent to when $ZnSO_4$ was the dietary zinc source.
Apparently the zinc of these water insoluble complexes dissoci-
ates in the digestive tract or the complex is not strong enough
to prevent the intestine from extracting the zinc. However, when
phytate is present in the diet in 12 to 15-fold molar excess over
zinc rats exhibit signs of decreased bioavailability of dietary
zinc (9,10). The metabolic balance does not measure true
absorption, but a measure can be obtained over time of the amount
of bioavailable nutrient under a given dietary regimen. Decreased
bioavailability would be indicated by a decrease in apparent

Table IV. Serum Zinc Response to Bran Muffins

Subject Subgroup	Day 1	Day 16	Day 31
		μg/100 ml	
A	92+23	(+)[2]117+16	(-)111+12
B	80+11[a]	(-) 136+8[b]	(+)110+13[c]

[1]Mean ± S.D., Fasting blood sample morning of day indicated.
[2](+)Whole bran muffins preceding 15 days, (-) dephytinized bran muffins preceeding 15 days.
a,b,c Significantly different, P < 0.05.

absorption, that is an increase in fecal excretion relative to intake. Apparent absorption of zinc by subjects on our study was lower the first 5 days the whole bran muffins were consumed than when dephytinized bran muffins were consumed, mean -0.2 and 1.5 mg/day, respectively. This indicates that the 12 molar ratio of phytate/zinc decreased bioavailability of zinc the initial 5 days whole bran muffins were consumed and evidently prevented reabsorption of zinc that had been secreted into the intestine (some individuals excreted more in the stool than they consumed). However, the relation reversed over the remaining 10 days each regimen was eaten, 3.1 vs. 2.4 mg/day for whole bran and dephytinized bran muffins, respectively. Reversal of the apparent absorption values suggest an adaptation to the diet, either a change in the ability of the intestine to absorb zinc or a homeostatic response which decreases secretion of endogenous zinc into the intestine.

The intake of zinc in our study was greater than from some hospital and self chosen diets (20,21,22) and exceeded slightly the recommended dietary intake (3). We did not determine the usual intake by our volunteers prior to the study and it is possible that the magnitude of the balances was influenced by the high intakes. The fasting serum zinc values during the study were markedly elevated compared to the initial value, supporting the conjecture that they consumed more zinc during than prior to the study. Prasad summarizes literature indicating that balance may increase during a period of adequate zinc intake when preceded by a period of low intake (23). Women showed increased plasma response to a zinc load following a vegetarian diet compared to after a diet that contained meat (24).

The dephytinized bran contained the same percentage of neutral detergent fiber as the nonincubated whole bran. Thus, the intakes of neutral detergent fiber did not differ between the two dietary regimens. Experiments on iron absorption by humans indicate the dephytinization process may alter binding of iron by dietary fiber in wheat bran (25). Lower zinc balances when dephytinized wheat bran was consumed might indicate the affinity for zinc by the fiber in dephytinized bran may be greater than that of the fiber in the unincubated whole bran. That conclusion is equivocal, however, if based only on our study.

The molar ratio of phytate/zinc of some mid-eastern diets was estimated to be 20 (26), but only small amounts of meat were generally available. Sandstrom et al. obtained results that indicate absorption of zinc from a meal may be appreciably influenced by the type of protein in the meal (27,28). Our menus provided protein from meat in most of the meals which may have affected bioutilization of the dietary zinc. In one comparison Sandstrom et al. found that whole wheat bread decreased percentage absorption, but the absolute amount of zinc absorbed was greater because total zinc in the whole wheat bread meal was greater.

Although the balance results indicate that in human diets a phytate/zinc molar ratio of 12 is not a deterrent to adequate zinc nutriture in adult men, qualifications as discussed above raise questions in interpreting the practical application. We have pursued two questions (1) What is the molar ratio of phytate/zinc in our typical diets? (2) What would be the effect of lower zinc intakes in a metabolic balance study with phytate/zinc molar ratio as the variable? Part of question 1 has been completed, but only preliminary information is available on question 2.

Analysis of Hospital and Self-Chosen Diets

Sample trays of three different hospital menus, regular, ovo-lacto vegetarian and soy-meat substitute, were collected for seven consective days and analyzed for minerals, phytic acid and fiber (29). Each meal was composited, homogenized and analyzed separately. The calculated caloric level of the meals was 2800 kcal per day.

The mean daily amount of zinc, phytate, fiber and the molar ratio phytate/zinc provided by the three diets is summarized in Table V. Only the soy-meat substitute diet supplied the recommended dietary intake of zinc for adult males. The values found for the regular and ovo-lacto vegetarian menus are similar to those reported by Brown et al. (20). The phytic acid in the ovo-lacto vegetarian menu was slightly greater than in the regular menu, but both were considerably less than provided by the soy-meat substitute menu. The phytate/zinc molar ratio of soy-meat substitute menu was about 2.3 and 1.7 times greater than of the regular and ovo-lacto vegetarian menus. The ratio of 7.6 is, however, considerably lower than 12 from our human study I. Neutral detergent fiber content of the regular and ovo-lacto vegetarian diets was significantly less than in the soy-meat substitute and the latter diet supplied an amount equal

Table V. Analysis of Zinc, Phytic Acid and Fiber Supplied by
Three Hospital Diets (26)[1]

Nutrient	Regular	Ovo-Lacto Vegetarian	Soy-Meat Substitute
Zinc mg/day	12.2 ± 3.3[a]	9.7 ± 1.4[a]	15.0 ± 1.8[b]
Phytic acid mg/day	387 ± 60[a]	440 ± 210[a]	1144 ± 144[b]
Phy/Zn[2]	3.3 ± 1.4[a]	4.5 ± 0.8[b]	7.6 ± 0.7[c]
Fiber[3] g/day	7.8 ± 0.6[a]	9.4 ± 3.3[a]	16.9 ± 1.7[b]

[1]Mean ± SD for 7 consecutive days. Means in a line with
different superscript are significantly different by Duncan's
multiple range test, $P < 0.05$.
[2]Molar ratio phytate/zinc.
[3]Neutral detergent fiber.

to our human study I. The slight increase in phytate and fiber
of the ovo-lacto vegetarian over the regular diet indicates that
the former probably contained more whole grain items than the
regular diet. But, not enough whole grain products were added
to the ovo-lacto vegetarian menu to compensate for the low zinc
concentration in dairy products substituted for the meat, and
the zinc supplied by this menu was low.

Prior to participating in a second balance study, described
in brief below, 15 volunteers maintained for 1 week, diet diaries
and brought to the laboratory 2nd plate collections duplicating
all food and liquid consumed. Food, including milk and fruit
juices, and liquids were kept separate. Analysis of the food
composites is completed. The bulk liquids will contribute little
additional zinc, therefore we know a close approximation of the
zinc intakes. The mean daily zinc content of the food composites
was 11.2 mg/day (SD of ± 3.9) and means for individuals ranged
from 5.6 to 20.6 mg/day. Individual daily intakes ranged from
2.1 to 29.2 mg. The mean phytic acid intakes for individuals
ranged from 300 to 2500 mg/day and the molar ratio of phytate/
zinc from 3.5 to 16.4. Only two of the 15 individuals consumed
diets with mean phytate/zinc molar ratio greater than 10. The
overall mean molar ratio of phytate/zinc was 7.2 ± 3.2.

Our analysis of hospital and self-chosen diets agrees with
information by Holden et al. and Freeland-Graves et al. (21,22)

that the majority of the United States population do not consume the recommended dietary intake of zinc. Additional information is being published to aid in estimating the phytate/zinc molar ratio of menus (26).

Metabolic Balance Study II

Three primary objectives were set to differentiate study II from study I; (1) a wider range of phytate/zinc molar ratio, (2) generally lower overall intake of mineral nutrients, and (3) to obtain information about dietary intakes and balances of the volunteers on their self-chosen diets prior to the controlled diet portion of the study.

Objective 3 is partially completed by the analysis of self-chosen diets discussed above. Excreta collections were conducted over the week the 2nd plate collections were made and when analyzed we will calculate balances to compare with the intakes and balances while consuming the controlled diet.

For the controlled diet portion of the study we again used menus of ordinary foods, but decreased the amount of meat and increased the amount of high carbohydrate foods. This attained some reducton in the zinc content of the menus. Whole wheat bran was not used in this study, instead, bran was dephytinized by endogenous phytase and only the water insoluble residue was used, baked into muffins. About 60% of the starting dry matter was recovered in the insoluble fraction and it contained no detectable phytate, 65 μg/g of zinc and 78% neutral detergent fiber. The water insoluble dephytinized bran was used at a level equivalent to about one half of the level of whole bran used in the previous study. Thus less bran, and bran of lower zinc content, enabled additional decrease in the total zinc intakes from the previous study. We added sodium phytate to the muffins as a source of dietary phytate. Three levels of phytic acid intakes were thus obtained, the dephytinized bran muffins without added sodium phytate and dephytinized bran muffins with two different levels of sodium phytate added. Analysis of the daily food composites indicate that the mean daily zinc intake will be about 10 to 12 mg per day, depending upon the caloric level. We did attain an intake more nearly equal to the usual mean intakes and the molar ratios of phytate/zinc will be about 4, 14 and 24. The treatment periods were again 15 days with 3 repeats of a 5-day menu cycle. Twelve volunteers consumed the three phytate levels in a random rotation. Urine and stools were collected and pooled over each 5-day menu cycle. We will be able to calculate balances for each 5-day period. Stools, but not urine, from one 5-day period have been analyzed and indicate that most balances will be positive, but smaller in magnitude than the previous study. Blood samples were obtained for serum zinc analysis at the initiation of the controlled diet period, at the end of each 15-day treatment period and 15 days post-study. The mean pre study value was about 20 µg per 100 ml greater than the previous study, Table VI. There was not, however, an increase during the controlled diet periods which we interpret as indicating the zinc intake matched closely their usual intakes. Phytate level did not affect serum zinc levels and there was no change 15 day post study.

Table VI. Serum Zinc Response to Three Levels of Phytate

	Dietary Treatment			
Pre[1]	Low[2]	Medium	High	Post[1]
		g/100 ml		
106	110	112	109	113

[1]Self-chosen diets.
[2]15 days each level of phytate.

Conclusion

Without the results from our second study, a definitive conclusion is not easy. If results of our 1st study are applicable to a range of dietary zinc intakes, a phytate to zinc molar ratio of 10 to 12 is not a hinderance to bioutilization of zinc by humans and the level of intake of zinc becomes a limiting factor in determining adequacy . We know from our analysis of self-chosen diets, and estimation by others, that most people consume diets that are less than 10 in phytate/zinc molar ratio. Vegetarians, particularly those who eat a lot of cereal products, will very likely exceed the value of 10. There is wide variation between individuals and from day to day by each individual. The role of adaptation or response of homestatic control mechanisms require further clarification but it is our feeling at present that humans can consume a wider molar ratio of phytate to zinc than heretofore thought.

Acknowledgment

The authors thank Priscilla Steel, Sheryl Cottrell and staff of the Human Study Facility, BHNRC: Dr. Phylis Moser and Taryn Moy, College of Human Ecology, University of Maryland, College Park, and David Hill, VMNL, BHNRC for their cooperative efforts toward the metabolic balance studies.

Literature Cited

1. Underwood, E.J. "Trace Elements in Human and Animal Nutrition" Academic Press, New York, 1977, p. 196.
2. Prasad, A.S. In: Harper, A.E.; Davis, G.K. "Nutrition in Health and Disease and International Development: Symposia from the XII International Congress of Nutrition", Alan R. Liss, Inc., New York, 1981; p. 165.
3. National Research Council "Recommended Dietary Allowances" National Academy of Sciences, Washington, DC, 1980.
4. Solomons, N.W. Am. J. Clin. Nutr, 1982, 35, 1048.
5. O'Dell, B.L.; Burpo, C.E.; Savage, J.E. J. Nutr. 1972, 102, 653.
6. Kelsay, J.L. Cereal Chem. 1981, 58, 2.
7. Reinhold, J.G.; Salvador, G.L.; Garzon, P. Am. J. Clin. Nutr. 1981, 34, 1384.
8. Davies, N.T.; Reid, H. Br. J. Nutr. 1979, 41, 579.
9. Davies, N.T.; Olpin, S.E. Br. J. Nutr. 1979, 41, 591.
10. Morris, E.R.; Ellis, R. J. Nutr. 1980, 110, 1037.
11. Davies, N.T.; Hristic, V.; Flett, A.A. Nutr. Rep. Int. 1977, 15, 207.
12. Morris, E.R.; Ellis, R. J. Nutr. 1980, 110, 2000.
13. Morris, E.R.; Ellis, R. Cereal Chem. 1981, 58, 363.
14. Ellis, R.; Morris, E.R.; Philpot, C. Anal. Biochem. 1977, 77, 536.
15. American Association of Cereal Chemists. "Approved Methods of the American Association of Cereal Chemists" American Association of Cereal Chemists, St. Paul, MN, 1977, #32-30.
16. Morris, E.R.; Ellis, R. Cereal Foods World 1979, 24, 461.
17. Morris, E.R.; Ellis, R.; Steele, P.; Moser, P. Fed Proc. 1980, 39, 787.
18. Morris, E.R., Ellis, R.; Steele, P.; Moser, P. in Hemphill, D.D. ed. "Proceedings 14th Annual Conference on Trace Substances in Environmental Health" Univ. of Missouri Press, Columbia, MO, 1981, p. 103.
19. Ellis, R.; Morris, E.R.; Hill, A.D. Nutr. Res., 1982, 2, 319.
20. Brown, E.D.; McGuckin, M.A.; Wilson, M.; Smith, J.C. J. Am. Diet. A. 1976, 69, 632.
21. Holden, J.M.; Wolf, W.R.; Mertz, W.M. J. Am. Diet. A., 1979, 75, 23.
22. Freeland-Graves, J.H.; Bodzy, P.W.; Eppright, M.A. J. Am. Diet. A. 1980, 77, 655.
23. Prasad, A.S. "Trace Elements and Iron in Human Metabolism" Plenum Medical Book, Co., New York 1978, p. 294.
24. Freeland-Graves, J.H.; Ebangit, M.L.; Hendrikson, P.J. Am. J. Clin. Nutr. 1980, 33, 1757.
25. Simpson, K.M.; Morris, E.R.; Cook, J.D. Am. J. Clin, Nutr., 1981, 34, 1469.

26. Oberleas, D.; Harland, B.F. J. Am. Diet. A. 1981, 79, 433.
27. Sandstrom, B.; Arvidsson, A.; Cederbald, A.;
    Bjorn-Rasmussuen, E. Am. J. Clin. Nutr., 1980, 33, 739.
28. Sandstrom, B.; Cederbald, A. Am. J. Clin. Nutr., 1980, 33,
    1778.
29. Ellis, R.; Morris, E.R.; Hill, A.D.; Smith, J.C. J. Am.
    Diet. A., 1982, 81, 26.

RECEIVED October 5, 1982

Zinc Bioavailability from Processed Soybean Products

JOHN W. ERDMAN, JR., RICHARD M. FORBES, and
HIROMICHI KONDO[1]

University of Illinois, Departments of Food Science and Animal Science,
Urbana, IL 61801

Soybean foods were utilized as models for the
evaluation of the effects of processing conditions
upon the relative bioavailability of both endogenous
and added zinc. Products such as full-fat soybean
flour, soy beverage, soy concentrates, soy isolates
and tofu were processed under carefully controlled
conditions and were individually incorporated into
diets for rat zinc bioassay studies. Some products
were tested under various in vitro conditions to
determine the factors that affect the binding of
zinc to components of the soy products. The overall
conclusions from these studies are that
phytate-to-zinc molar ratios alone do not predict
zinc bioavailability and food processing conditions
markedly affect the bioavailability of zinc
endogenous to soy foods. Inorganic zinc ($ZnCO_3$)
added (fortified) to soy product-containing diets
was highly available.

Often overlooked in the evaluation of the effects of diet upon
mineral availability is the role that food processing plays in the
formation of or breaking of ligand-metal complexes. Several
individual or unit processing steps are needed to produce a soy
concentrate, a bread or a spray-dried egg white. Some or all steps
may have a bearing upon final mineral bioavailability. Soy
concentrate from company A is not produced in precisely the same
manner as from company B. In fact, lot to lot variation for the
same product may be quite variable, particularly in mineral
content.
  Rackis and coworkers (1,2) noted from their survey of the
literature that for experimental animals, zinc bioavailability from
soy protein isolates was low compared to zinc availability from
soybean meal, casein or other animal protein diets. They

[1] Current address: Kinjo Gakuin University, Nagoya, Japan

attributed the differences in availability to formation of
phytate-protein-mineral complexes during the processing of
isolates. Other researchers, notably Lease (3,4), O'Dell (5,6) and
Forbes (7,8) have reported experimental results that indicated the
variable nature of zinc bioavailability from soy protein products.
Oberleas (9) first suggested that the molar ratio of phytate
to zinc might be useful for prediction of the zinc bioavailability
from phytate-rich foods. Molar ratios of greater than 20:1 seemed
to be indicative of poorly available zinc. O'Dell (10), Morris and
Ellis (11) and Davies and Olpin (12) have all pointed out the
importance of the calcium content of the diet to the phytate to
zinc molar ratio. Higher dietary calcium clearly depresses zinc
bioavailability at phytate to zinc molar ratios of less than 20:1
in diets fed to rats.
Use of the phytate to zinc molar ratio may be too simplistic
for prediction of zinc availability from mixed food systems.
Besides the effect of dietary calcium level, the amount of iron,
and perhaps the level of other metals, may affect zinc
bioavailability. Solomons and Jacob (13) have shown in human
subjects that increasing the iron/zinc ratio from 0:1 to 3:1 in
solutions containing 25 mg of zinc and corresponding amounts of
iron as ferrous sulfate produced a progressive decrease in the
plasma zinc response. They further reported that the chemical form
of iron was an important determinant of the interaction. Solomons
(14) extensively reviewed both inhibitory factors and enhancers of
zinc bioavailability found in foods.
Our laboratories have been concerned with the role that unit
food processing operations play in the bioavailability of zinc from
complete diets. Soybean foods have served as models for the
evaluation of processing effects upon both endogenous and added
zinc. Below are described results from both rat bioassays and in
vitro tests for zinc bioavailability. Prediction of zinc
bioavailability from soy-containing diets is far more complex than
an analysis of phytate and zinc molar ratios.

Methods

For most of the in vivo tests of zinc bioavailability reported
here, the slope ratio technique suggested by Shah and coworkers
(15) and further developed in our laboratories (16,17) was
utilized. Briefly, isonitrogenous, isoenergetic 20% protein diets
are fed to groups of weanling rats for 21 days. Control diets
containing 20% egg white protein are supplemented with one of
several levels of zinc as zinc carbonate. To test for the
availability of zinc in a test soy protein product, several levels
of zinc are added to egg white-based diets by substituting the soy
product for egg-white on an equivalent protein basis. To test for
the effect of the presence of a soy protein product in a diet upon
the bioavailability of zinc added (fortified) to diets, zinc
carbonate is added at various levels to diets containing a specific

mixture of soy protein and egg white. After 21 days, rats are killed and either tibias or femurs are removed. To provide a measure of relative zinc availability (15-17), the data for the weight gain and for the total bone zinc are statistically analyzed by regression analysis to compare the slopes of the linear portions of the lines relating response per unit of added mineral.

For in vitro tests the method developed by Miller and coworkers (18) to estimate iron availability from meals was adapted for zinc testing. We have also examined the solubility of zinc in aqueous extracts of soy products by the methods published by de Rham and Jost (19).

Results and Discussion

Bioavailability of Zinc Intrinsic to Soybean Products and Added to Soybean Products. A number of studies (16,17,20) was

Table I

Relative Bioavailability of Zinc Intrinsic to Soy Products

Product	Phytate-to-zinc molar ratio	Relative Bioavailability (%)[a]	
		Weight gain	Log bone zinc
Full-fat soy flour	28	55[b]	34[b]
Freeze-dried soy beverage	26	63[b]	40[b]
Spray-dried soy concentrate (neutral form)	52	41[b]	20[b]
Freeze-dried soy concentrate (neutral form)	57	66[b]	29[b]
Freeze-dried soy isolate (neutral form)	34	85	46[b]
Freeze-dried soy concentrate (acid form)	66	113,93[c]	48[b]
Freeze-dried soy isolate (acid form)	35	106	64[b]

[a]Data points represent comparisons of slopes of responses of test diets (containing soy) with responses of control diets (without soy products) with added minerals as the carbonate X 100. (Sources: 16,17,20).
[b]Significantly different from control diets (P < 0.05).
[c]Experiment was run twice.

performed to investigate the bioavailability to the rat of zinc
intrinsic to various soybean products (Table I). The relative
bioavailability of zinc from soy products was highly variable but
usually low when compared to the availability of zinc from zinc
carbonate. · The phytate-to-zinc molar ratio was a poor predictor of
zinc bioavailability.

The acid forms of the isolates and concentrates demonstrated
excellent bioavailability for zinc relative to neutralized products
prepared under identical conditions (20). The difference may be
due to the formation of stable protein-phytic acid-zinc complexes
in the dried neutral product. Protein-phytic acid-mineral
associations have been shown to occur in solution at a neutral pH
(19). These associations may well form more tightly-bound
complexes during the drying of the soy protein. The exclusion of
water from the protein-phytic acid-zinc associations could lead to
thermodynamically stable complexes that are resistant to complete
proteolytic digestion in the gastrointestinal tract. Short
peptides or amino acid residues bound to zinc and phytic acid then
would be poorly absorbed. For further discussion see Erdman et al.
(20) and Cheryan (21).

Results from studies investigating the effect of the presence
of soy products upon the bioavailability of exogenous (fortified)
zinc, added as the carbonate are shown in Table II. The results
demonstrate that the presence of soybean products in complete diets

Table II

Effect of Soy Products Upon the Relative Bioavailability
of Extrinsic Zinc Added as Zinc Carbonate

Product	Phytate-to-zinc molar ratio	Relative Bioavailability (%)[a]	
		Weight gain	Log bone zinc
Full-fat soy Flour	28	94	94
Spray-dried soy concentrate (neutral form)	52	77[b]	70[b]
Freeze-dried soy concentrate (neutral form)	57	84	78
Freeze-dried soy concentrate (acid form)	66	65	83

[a]Data points represent comparisons of slopes of responses of test
diets (containing soy) with responses of control diets (without
soy products) with added minerals as their carbonate X 100.
(Sources: 16,17,20).
[b]Significantly different from control diets ($P < 0.05$).

has little detrimental effect upon the bioavailability of added
zinc. Other workers, including Hardie-Muncy and Rasmussen (22),
have also reported that inorganic zinc added to soy isolates to be
better utilized by rats than zinc intrinsic to the isolate.
Therefore, these studies suggest that fortified zinc in soy
products is of high bioavailability relative to intrinsic zinc.

Recently, our laboratories and Weaver and coworkers from
Purdue University have begun a collaborative effort to investigate
the bioavailability of zinc from various soy products utilizing
intrinsically-labeled and extrinsically-labeled [65]zinc soy
products. The [65]zinc intrinsically labeled soybeans were
obtained by growing Century variety soybeans hydroponically as
described by Levine et al. (23). [65]Zinc intrinsically labeled
soybeans were defatted. Acid and neutralized soy concentrates were
prepared as previously described (20). Preliminary results of a
rat feeding study are shown in Table III (Ketelsen, et al.,
manuscript in prep.). Groups of male weanling rats were fed
experimental unlabeled diets containing 10 ppm zinc from defatted
soy flour, acid-precipitated soy concentrate or neutralized soy
concentrate for 7 days. On the evening of the seventh day, each
rat was fasted and then given a [65]zinc intrinsically-labeled
test meal which was similar to the diet fed for the last seven
days. After the meal the rats were individually placed into a
whole body gamma counting chamber and the [65]zinc activity was
recorded and designated as the day 0 activity. Rats were then
refed the same unlabeled experimental diets as had been fed prior
to the test meal. Whole body [65]zinc activity was measured 24
hours and 12 days after administration of the test meal. Table III
shows the percent whole body retention from the three test meals
after 1 and 12 days. The results verify results from Table I,
demonstrating the variable bioavailability of zinc from different
soy products. The retention of zinc from the meal containing the
acid-precipitated soy concentrate was greater than from the neutral

Table III
Whole Body Retention of [65]Zinc from Defatted Soy Flour, Acid
Precipitated- and Neutralized Soy Concentrate-Based Meals (%)[a]

Diet	Day 1 (mean ± S.D.)	Day 12 (mean ± S.D.)
Defatted soy flour	83.1 ± 4.1[a]	75.5 ± 4.0[a]
Acid concentrate	75.9 ± 5.5[b]	68.0 ± 5.0[b]
Neutral concentrate	64.7 ± 4.4[c]	52.3 ± 4.4[c]

[a]N = 6 or 7. In vertical columns, means not sharing a common
superscript letter are significantly different (P < 0.05).
(Source: Keteleson et al., manuscript in prep.).

product, again suggesting that neutralization of soy protein results in formation of protein-phytate-mineral complexes.

As pointed out earlier in this review, increasing the level of dietary calcium decreases the zinc bioavailability from phytate-containing foods. Presumably the mechanism is through the formation of chemical complexes containing zinc, phytate and calcium which are insoluble at intestinal pH and nonabsorbable (24). Recently, our laboratories used slope ratio techniques to compare the bioavailability of zinc contained in calcium sulfate- and in magnesium chloride-precipitated soybean curd (Tofu) to that of zinc added as the carbonate to egg white diets by slope ratio techniques (25). Total dietary calcium level in all diets was adjusted to 0.7% with calcium carbonate. The results (not shown) indicated that the relative availability of zinc from both tofu preparations was 51% as measured by weight gain and 36-39% for bone zinc. These results are similar to those reported for full fat soy flour (16) in Table I.

In a separate experiment with zinc supplied at 9 ppm in all diets, the effect of increasing dietary calcium in calcium or magnesium precipitated tofu or egg white diets on weight gain and tibia zinc accumulation was tested. From Table IV it can be noted that the performance of tofu-fed rats relative to zinc carbonate-fed rats was quite similar at 0.4% total dietary calcium, but was reduced as dietary calcium was increased to 0.7% and to 1.2%. These results suggest that a poorly available calcium-zinc-phytate complex not present in the soy curd can form in the gastrointestinal tract when sufficient calcium is added to the diet.

Table IV

Effect of Dietary Calcium Level Upon Weight Gain and Tibia Zinc of Rats fed 9 ppm Zinc from Calcium or Magnesium Precipitated Tofu or from Zinc Carbonate Added to Egg White Diets

	Dietary Calcium Level					
	0.4%		0.7%		1.4%	
Diet	Wt. gain (g)	Tibia Zinc (μg/2 tibia)	Wt. gain (g)	Tibia Zinc (μg/2 tibia)	Wt. gain (g)	Tibia Zinc (μg/2 tibia)
Egg	117 + 5	50 + 2[a]	105 + 8[a]	46 + 1[a]	95 + 11[a]	45 + 2[a]
Ca-Tofu	112 + 7	34 + 2[b]	75 + 5[b]	25 + 1[b]	46 + 5[b]	24 + 4[b]
Mg-Tofu	114 + 5	37 + 2[b]	70 + 8[b]	28 + 1[b]	46 + 3[b]	25 + 1[b]

[a]Means + SEM. For columns, means not sharing a common superscript letter are different (P < 0.05) (Source: 25).

Calcium Effects on Zinc Bioavailability for the Rat and the
Human.   It should be pointed out at this juncture that the nutrient
requirement of calcium for the rat is much higher than for man.   In
fact, the molar ratio of calcium to zinc in excess of 660:1 is
recommended for rat diets, while for man the ratio is between 80:1
and 160:1.   To feed rats molar ratios of calcium and zinc similar
to human requirements would necessitate either a very calcium
deficient diet or one containing zinc at a level well in excess of
the requirement.   Neither choice is nutritionally suitable for
demonstrating an effect of phytate on zinc availability.
     The calcium tofus prepared for the study found in Table IV
contained a calcium-to-zinc molar ratio of 330:1 to 355:1.   This
ratio is higher than would be attained in a balanced diet for
humans, but it is lower than the 700:1 ratio present in the 0.4%
calcium diet fed to the rats.   Since this latter ratio had a
minimal effect on zinc availability in rats one would expect the
zinc from a meal of calcium tofu to be well utilized by humans.
This assumes, of course, that the rat and the human respond
similarly to these molar ratios.

Table V

In Vitro Availability of Zinc from Various Soy Products[a]

Soy Product	Phytate to zinc molar ratio	Calcium to zinc molar ratio	Dialyzability (%)
Mature soybean	35	73	26.71
Full-fat soy flour	29	81	25.13
Dehulled soybean flour	34	88	11.03
Green soybean	18	150	9.51
Soy beverage	25	86	6.44
Spray-dried, neutralized soy concentrate	52	90	4.70
Low phytate soy concentrate	13	440	3.06
Normal phytate soy concentrate	44	156	4.12
Acid concentrate	35	87	1.20
Neutral concentrate	33	83	0.60
Calcium tofu	28	359	1.32
Magnesium tofu	26	38	0.85

[a]Method: D. D. Miller et al. (18).   Unpublished data of Kondo
     et al.

We would have to question the validity of the use of dietary calcium levels of up to 1.75% calcium in rat studies to study zinc bioavailability. These levels would never be approached in human diets without use of excessive calcium supplementation.

In Vitro Tests for Zinc Bioavailability. In vitro digestibility testing can yield valuable information regarding the solubility of chelates from foods. The extent of dialysis of zinc from a partially or fully digested test meal can demonstrate the presence of chelates, suggest the molecular size of chelate complexes and is predictive of bioavailability of minerals. The method of Miller et al. (18) was adapted to determine the % dialyzability (in vitro bioavailability) of zinc from several soy products (Table V). Many of the products with the highest dialyzability are the least refined. Some of the concentrates, as well as the tofu products, have very low dialyzability. The results of these in vitro tests do not necessarily agree with the results of the in vivo rat bioassays.

It is noteworthy that the phytate-to-zinc molar ratio does not predict in vitro dialyzability by this procedure. The calcium plus magnesium-to-zinc molar ratio may be more predictive of in vitro dialyzability. It can be seen in Table V that all soy products with calcium-to-zinc molar ratios of greater than 100 fall below 10% dialyzability. Of special note is the magnesium tofu product. It has an extremely low calcium-to-zinc molar ratio (38) but the highest (217) magnesium to zinc ratio and a low in vitro availability. As will be seen in Table VI, magnesium aggravates loss of zinc dialyzability almost to the same extent as does calcium.

Table VI
Effect of Calcium and Magnesium on In Vitro Bioavailability
of Zinc from Defatted Soy Flour[a]

Dialyzability of Soy Zinc (%)

Calcium (mg) added as $CaCO_3$[b]

Added Ca	0	10	25	50	100	200	300	600
%	17	12	7.5	5	5	5	4.5	5

Magnesium (mg) added as $MgCO_3$[b]

Added Mg	0	10	25	50	100	200	400
%	17	14.5	10.5	8.5	7	5.5	5.5

[a]Method: Miller et al. (18). Unpublished data of Kondo et al.
[b]Added to 12.5 gm of defatted flour.

Although, the modified Miller procedure did not rank soybean products similarly to rat bioassay, when the procedure was used to test for the effect of adding calcium to soy flour, there was a clear demonstration of the inhibitory effect of calcium addition on dialyzability of zinc from soy flour (Table VI). Magnesium ion also reduced in vitro bioavailability of zinc. The effect of magnesium was noted previously in vitro (19), but has not been adequately tested in vivo. Iron added as ferrous ion at similar levels did not affect zinc dialyzability (data not shown).

In vitro tests using other procedures (19) were carried out in our laboratory (Ketelsen et al., manuscript in prep.) to study the dialyzability of zinc from slurries of defatted soy flours over a range of pH's. Either ^{65}zinc was added to soy slurries as ^{65}zinc carbonate (extrinsic label) or the ^{65}zinc was intrinsically labeled in soybeans grown hydroponically. Results from these studies (data not shown) indicated that the extrinsic and intrinsic zinc pools do not completely mix. The results showed that the pH of the soy suspensions affected the quantity of zinc that was bound in complexes of molecular weight in excess of 12,000. Also, intrinsic zinc was more likely to be bound to high molecular weight complexes than extrinsic zinc, especially for slurries at pH 3.

Table VII
Factors Affecting the Bioavailability of Zinc from Soy Products

---

Nutritional status of man (animal)
Dietary phytic acid level
Dietary calcium level
Source of zinc (intrinsic/extrinsic)
Unit food processing operations
   -pH adjustment
   -level of refinement
   -addition or removal of inhibitors/enhancers
   -other factors
Digestability of product
Dietary level of magnesium, iron, etc.
Stage of maturity of bean
Others

Our work with soy products indicates that extrinsic zinc is more available than intrinsic zinc. The intrinsic and extrinsic zinc pools do not completely mix in all soy products. Phytic acid is an inhibitor of zinc bioavailability and this inhibition is aggravated by higher levels of dietary calcium and perhaps magnesium. Neutralization of soy isolates and concentates, with subsequent drying, reduces zinc utilization for rats.

Conclusions

The prediction of zinc bioavailability from complex food systems is not a simple matter. Animal bioassays and in vitro tests help us to identify factors that may enhance or inhibit zinc utilization from the diet. With simple model food systems, we can demonstrate the negative effects of phytic acid, calcium and other factors on zinc bioavailability. However, the interaction of these factors in complex food systems, and their effect on zinc status for man is not well understood at this time.

Simple predictors, such as the phytate to zinc molar ratio, will not be accurate for human diets. The aggravating effect of calcium and perhaps of magnesium seen in rat and in vivo tests may or may not be of practical significance to man. We can, however, develop a list of factors that have been shown to affect zinc bioavailability from foods. These are listed in Table VII.

Our work with soy products indicates that extrinsic zinc is more available than intrinsic zinc. The intrinsic and extrinsic zinc pools do not completely mix in all soy products. Phytic acid is an inhibitor of zinc bioavailability and this inhibition is aggravated by higher levels of dietary calcium and perhaps magnesium. Neutralization of soy isolates and concentrates, with subsequent drying, reduces zinc utilization for rats.

Acknowledgments

Funding for this work was provided, in part, by the American Soybean Assn. (ASARP #81602) and the USDA (CSRS Grant #616-15-172 under PL89-106).

Literature Cited

1. Rackis, J. J.; McGhee, J. E.; Honig, D. H.; Booth, A. N. J. Am. Oil Chem. Soc. 1975, 75, 249A-253A.
2. Rackis, J. J.; Anderson, R. L. Food Prod. Dev. 1977, 11, 38-44.
3. Lease, J. G.; Williams, W. P. Poultry Sci. 1967, 46, 233-242.
4. Lease, J. G. J. Nutr. 1967, 93, 523-532.
5. O'Dell, B. L.; Burpo, C. E.; Savage, J. E. J. Nutr. 1972, 102, 653-660.
6. Oberleas, D.; Muhrer, M. E.; O'Dell, B. L. J. Nutr. 1966, 90, 56-62.
7. Forbes, R. M.; Yoke, M. J. Nutr. 1960, 70, 53-57.
8. Forbes, R. M. Fed. Proc. 1960, 19, 643-647.
9. Oberleas, D. "Proceedings of Western Hemisphere Nurition Congress"; White, P.L; Selvey, N., Ed. Am. Med. Assn. Chicago, IL, 1975; 156.
10. O'Dell, B. L. "Soy Protein and Human Nutrition"; Wilcke, H. L.; Hopkins, D. T.; Waggle, D. H., Ed. Academic Press, New York, NY, 1979; 187.

11. Morris, E. R.; Ellis, R. J. Nutr. 1980, 110, 1037-1045.
12. Davies, N. T.; Olpin, S. E. Brit. J. Nutr. 1979, 41, 590-603.
13. Solomons, N. W.; Jacob, R. A. Am. J. Clin. Nutr. 1981, 34, 475-482.
14. Solomons, N. W. Am. J. Clin. Nutr. 1982, 35, 1048-1075.
15. Momcilovic, M.; Belonje, B.; Giroux, A.; Shah, B. G. Nutr. Rept. Int. 1975, 12, 197-203.
16. Forbes, R. M.; Parker, H. M. Nutr. Rept. Int. 1977, 15, 681-688.
17. Forbes, R. M.; Weingartner, K. E.; Parker, H. M.; Bell, R. R.; Erdman, J. W., Jr.; J. Nutr. 1979, 109, 1652-1660.
18. Miller, D. D.; Schricker, B. R.; Rasmussen, R. R.; Van Campen, D. Am. J. Clin. Nutr. 1981, 34, 2248-2256.
19. de Rham, O.; Jost, T. J. Food Sci. 1979, 44, 596-600.
20. Erdman, J. W., Jr.; Weingartner, K. E.; Mustakas, G. C.; Schmutz, R. D.; Parker, H. M.; Forbes, R. M. J. Food Sci. 1980, 45, 1193-1199.
21. Cheryan, M. "CRC Crit. Rev. Food Sci. Nutr." 1980, 13, 297-335.
22. Hardie-Muncy, D. A.; Rasmussen, A. I. J. Nutr. 1979, 109, 321-329.
23. Levine, S. E.; Weaver, C. M.; Kirleis, A. W. J. Food Sci. 1982, 47, 1283-1287.
24. Oberleas, D.; Muhrer, M. E.; O'Dell, B. L. J. Nutr. 1966, 90, 56-62.
25. Forbes, R. M.; Erdman, J. W., Jr.; Parker, H. M.; Kondo, H.; Ketelsen, S. M. J. Nutr. (manuscript submitted).

RECEIVED September 3, 1982

Zinc Bioavailability from Cereal-Based Foods

G. S. RANHOTRA and J. A. GELROTH

American Institute of Baking, Nutrition Research, Manhattan, KS 66502

Under the current dietary practices, the intake
levels of phytate, fiber, plant proteins, etc.,
which may possibly inhibit zinc absorption, are not
a source of undue concern for the majority of the
population. However, as our awareness of the concepts of nutrient economics and nutrient density,
and of matters of health as they might relate to
foods of plant origin increases, more of the zinc
in our diet could originate from cereal-based and
other plant foods. Consequently, the total and bioavailable zinc content of our diet may become com-
promised. This may have serious health implications
especially when a steady decline in our caloric intake is being witnessed. Health measures encompassing fortification of foods with zinc, the processing of foods to control the levels of suspected
inhibitors of zinc absorption, and long-term studies
to examine the effect of "adaptation" (to changed
dietary regimens) are some of the areas which demand
our immediate attention.

Zinc deficiency in man was first recognized in the early
1960's (1). Since that time, research concerning the role of zinc
in health and disease has been intense. Evidence that a sizable
segment of the U.S. population may be deficient in zinc, at least
marginally (2), has contributed to the interest in zinc research.
Although the major sources of zinc in the diet are foods of
animal origin, particularly meats, shellfishes and cheese (3,4),
cereal-based foods also make important contributions.

Cereal-Based Food Products
Zinc content.
Table I lists the zinc content of some cereal-based food products (3,4). Substantial zinc loss occurs during refinement of cereal grains, e.g., milling of wheat to produce white flour. Therefore, products which are less refined or which contain the bran

and germ fractions of grains are particularly good sources of
zinc.

Table I. Zinc content of some cereal-based foods[a]

Product	mg/100 g
Cake, white	0.2
Sweet corn, boiled	0.4
Hamburger roll	0.6
White bread	0.6
Oatmeal bread	1.0
Graham crackers	1.1
Oatmeal cookies	1.3
Cake, chocolate	1.3
White rice, dry	1.3
Brown rice, dry	1.8
Macaroni, dry	1.8
Wholewheat bread	1.8
Breakfast cereals	
shredded wheat	2.8
bran flakes, 40%	3.6
germ, toasted	15.4

[a] Murphy et al. (3) and Freeland and
Cousins (4)

Zinc fortification.
    The current cereal fortification (enrichment) practice re-
stores some of the nutrients (zinc not included) lost during milling. In 1974, however, the Food and Nutrition Board (5) proposed
an expansion of the current program to include a total of 10 micronutrients (zinc included). The proposed level of zinc (Table II)
would almost completely restore zinc to the level present in whole
grains.
    Although great changes in dietary habits are currently being
witnessed, refined products are likely to remain the mainstay of
our diet. This and the fact that our caloric intake has been de-
clining in recent years, necessitates that the dietary adequacy
of certain nutritionally critical nutrients such as zinc be ensur-
ed. Fortification provides a mean to accomplish this.
    Zinc fortification sources.
    Should the nutritional and other considerations permit forti-
fication, a number of zinc sources can be used to fortify grain
products. These sources, unlike potential iron sources, appear not
to differ appreciably in their relative biological value(6). This
was observed in rats fed submarginal (9.5 ppm) zinc diets (Table
III). Femur zinc content was used to assess biological values
(BVs). Indirectly, this means that product compatibility and cost
are likely to be more important than nutritional considerations in
choosing the zinc source to be used in cereal fortification.

Table II. Expanded cereal fortification program[a,b]

	Fortification(enrichment)	
	Current (mg/100 g)	Expanded(1974) (mg/100 g)
Vitamin A (ret.eq.)	–	0.29
Thiamin	0.64	0.64
Riboflavin	0.40	0.40
Niacin	5.29	5.29
Vitamin B-6	–	0.44
Folic acid	–	0.07
Iron[c]	2.86–3.63	8.81
Calcium	211 (optional)	198.20
Magnesium	–	44.04
Zinc	–	2.20

[a] Food and Nutrition Board (5)
[b] Wheat flour, corn grits, corn meal and rice
[c] Level proposed (FDA) in 1981: 4.41 mg/ 100 g
(level proposed in 1974 was disallowed)

Zinc Bioavailability

Dietary adequacy of zinc is strongly linked to zinc bioavailability, particularly in foods representing the bread and cereal group. Cereal products currently provide about one-fifth of our energy need (7). Various dietary guidelines proposed in recent years to improve health have recommended that we increase our consumption of cereal foods to include more complex carbohydrates in the diet. These and other considerations have led to an impressive increase in our consumption of some grain products such as variety breads (8). This means more and more of the zinc in our diet originates from foods which also contain naturally occurring or added components suspected to inhibit the absorption of zinc. The components most widely investigated and which will be discussed include (a) phytates, (b) fiber, (c) protein and (d) certain micronutrients. Other factors (Table IV) are also important but will not be discussed.

Effect of phytates.

A number of studies have shown that zinc, as compared to other minerals, is poorly utilized from phytate-containing foods. The relative BV of zinc in a number of phytate-containing foods was recently examined by Franz et al (9). Each product was prepared as for human consumption and incorporated into a semi-purified diet fed to rats. Whole corn and brown rice had a low relative BV (0.58 or less compared to 1.00 for zinc in zinc sulfate) while wholewheat flour and unleavened wholewheat bread had medium values (0.63–0.74). Refined cereal products such as white flour, leavened and unleavened white bread and white rice (all low in phytate) had high relative BV (0.89–1.08) as did leavened wholewheat bread

Table III. Bioavailability of zinc in potential fortification sources

	Source (diet)						
	Carbonate	Chloride	Oxide	Sulfate	Stearate	Acetate	Elemental
Zinc content source (%)	53.8	44.7	79.5	36.1	11.2	30.0	100.0
diet (ppm)	9.5	9.5	9.5	9.5	9.5	9.5	9.5
Serum zinc (μg/dl)	167±18	154±12	165±19	140±24	144±20	144±18	161±11
Femur zinc (μg)	31.8±2.5	29.6±3.5	34.4±4.1	34.2±2.8	35.0±2.9	35.8±3.4	33.0±2.6
Zinc intake (μg)	694±92	766±36	775±58	730±43	751±57	729±58	698±97
Zinc absorbed (%)	76±5	89±3	87±4	87±4	89±2	87±5	88±4
RBV of zinc[b]	93	87	101	100	102	105	95

a Ranhotra et al. (6)
b Relative (zinc in zinc sulfate = 100%) biological value based on femur zinc.

Table IV. Factors affecting the bioavailability of zinc in food

-Form and chemical nature of zinc
-Dietary interactions (zinc vs. protein, lipids, carbohydrates such as fiber, other minerals,
 vitamins and chelators such as phytates)
-Food processing conditions (leavening, etc.)
-Digestibility of food
-Body's need for zinc/effect of "adaptation"
-Health status of the individual
-Intestinal microflora
-Effect of age, sex, medication, etc.
-Environmental and other factors

(1.04). The authors conclude that relative BV appeared to be inversely related to phytate contents of the foods. Another study (10) tested cookies made with egg albumin with phytate added or not. The addition of phytate (Table V) significantly reduced the bioavailability of zinc as determined based on the zinc content of femur (Table VI). The reduced BV was comparable to BVs obtained with cookies made with the soy protein products. In these studies, the level of protein and zinc in all test products was equalized. In the albumin-based cookies, however, the zinc originated almost entirely from the added source (zinc chloride), while in the soy-based cookies the added source made only a partial contribution. In these studies, very little phytate was hydrolyzed during the

Table V. Cookie formula[a]

	Cookie				
	A	B	C	D	E
Cake flour (g)	100	100	100	100	100
Other ingredients(g)	58	58	58	58	58
Protein source[b](g)					
egg albumin	33	33	–	–	–
soy flour (def.)	–	–	50	–	–
soy concentrate	–	–	–	39.9	–
soy isolate	–	–	–	–	30.9
Zinc (mg)					
added (as chlo.)	4.49	4.49	2.20	3.40	2.24
in protein source	0.07	0.07	2.36	1.16	2.32
Phytate (mg)					
added (as Na salt)	–	1107	–	66	448
in protein source	–	–	1107	1041	659
Phyt. hydrolyzed[c](%)	100	19.6	7.3	9.8	18.4

[a] Ranhotra et al. (10)
[b] Amount added provided 26 g protein
[c] In cookie-making

cookie-making process (Table V). In cereal-based products where appreciable phytate hydrolysis occurs during processing, bioavailability of zinc seems to improve. This was observed in a study (11) where breads, made with or without soy (Table VII), contained added phytate. In these breads, phytate was almost completely hydrolyzed during processing. The breadmaking process also significantly improved the apparent absorption of zinc (Table VII). Reinhold et al. (12) studied the availability of zinc in leavened and unleavened wholemeal wheaten breads and concluded that leavening (hydrolyzes phytate) markedly increased the bioutilization of zinc in bread. Morris and Ellis (13) examined the effect of dephytinizing wheat bran on zinc utilization by rats and demonstrated that it is feasible to greatly overcome the phytate effect through dephytinization (Table VIII). In humans, Sandstrom et al. (14) reported reduction in the availability of zinc added to wholemeal

Table VI. Tissue concentration and absorption of zinc in rats fed cookies[a]

| | Cookies (Table V) | | | | |
	A	B	C	D	E
Dietary zinc (ppm)	9.0	9.0	9.0	9.0	9.0
Body weight gain (g)	163±10	160±9	165±9	166±14	165±12
Serum zinc (µg/dl)	98±14	90±14	100±11	87±15	86±13
Femur zinc (µg)	45.5±2.8	33.6±2.9	34.2±2.7	33.4±3.0	32.4±4.1
Zinc intake (µg)	3197±162	2995±296	3188±182	3204±258	3132±296
Zinc absorbed (%)	89±2	85±2	83±3	83±3	81±3
RBV[b]	100	76	77	77	73

a Ranhotra et al. (10)
b Relative (zinc in cookie A=100%) biological value based on femur zinc content.

Table VII. Tissue concentration and absorption of zinc in rats fed bread[a,b]

| | Phytate added[c] | | | |
| | No soy | | Soy added (12%) | |
	Ingredients	Bread	Ingredients	Bread
Dietary zinc (ppm)	9.5	9.5	9.5	9.5
Body weight gain (g)	188±12	185±11	190±11	189±11
Serum zinc (µg/dl)	106±32	105±10	109±43	174±26
Femur zinc (µg)	53.4±6.3	57.6±4.9	50.1±5.9	55.7±6.5
Zinc intake (µg)	1168±64	1168±87	1242±64	1234±84
Zinc absorbed (%)	61±9	82±3	65±6	80±3
Phytate hydrolyzed (%)	-	100	-	92.2

a Ranhotra et al. (11)
b Flour contained 2.2 mg added zinc/100 g; zinc addition adjusted for zinc in soy.
c As sodium salt. Phytate added to provide 77 mg P (phytate in soy considered).

bread compared with white bread. But they also found that the "absolute" absorption of zinc naturally-occurring in wholemeal bread is high. This may also be true for all grain products which are good sources of zinc, unless these products have "inhibitory" (on zinc absorption) influences too great to be overcome during digestion and/or absorption processes.

Added and naturally-occurring phytates may affect zinc absorption differently although Morris and Ellis (13) report that phytate in wheat bran affects zinc utilization in rats in nearly the same manner as does phytate added at a similar phytate : zinc molar ratio. One can arrive at a similar conclusion in examining the data in Table VI where cookie B containing added phytate compared well with cookie C which contained naturally-occurring phytate.

Phytate : zinc molar ratios in food products can be used to estimate the relative risk of having an inadequate intake of zinc. Oberleas and Harland (2) propose that a daily phytate : zinc molar ratio of 10 or less is acceptable in providing adequate zinc and that daily ratios above 20 may jeopardize zinc status.

Effect of fiber.
Unlike phytates, the effect of fiber on zinc absorption remains more uncertain. The information recently compiled (Table IX) by Kelsay (15) underscores this. The uncertainty may, in part, be attributed to (a) differences in preexperimental dietary regimen of test subjects, (b) differences in the level of fiber intake, (c) differences in the type of fiber tested, (d) differences in the time span of the study (effect of "adaptation"), and most critically (e) the experimental difficulties in disassociating the effect of other suspected inhibitors from effect due to fiber. Conceivably, the nature of the zinc complex (involving fiber, phytate, protein, etc.) existent at the site of zinc absorption may be more critical than the mere levels of potential inhibitors.

The effect of "adaptation" may be particularly significant. Anderson et al. (16) studied the zinc status of long-term vegetarian women who obtained 77% of the total dietary zinc from plant products. Three-day dietary records showed that these subjects consumed 9.2 ± 2.5 mg zinc and 30.9 ± 11.0 g dietary fiber originating from five different food groups. Their zinc status (serum zinc (μg/dl): 98.3 ± 23.9; hair zinc (μg/g): 187 ± 44) appeared within the normal range despite their high intake of less rapidly absorbed zinc and their high intake of total fiber and phytate.

Table VIII. Femur zinc concentration in rats[a]

Dietary zinc source	Femur zinc (ppm)
Zinc sulfate	177+10
Raw wheat bran	69+3
Low-phytate bran	136+6

[a] Morris and Ellis (13). Dietary zinc: 12 ppm

Effect of protein.
Many earlier studies have shown that zinc is less efficiently

Table IX. Effect of fiber on zinc balance[a]

Investigator	Fiber source	Zinc balance
Reinhold et al. (1973)	Unleavened wholemeal bread (350 g/day for 32 days)	Became negative with wholemeal bread
Reinhold et al. (1976)	Leavened wholemeal bread (60% of calories) for 20 days	Became negative with wholemeal bread
Ismail-Beigi et al. (1977)	Cellulose (10 g) in apple compote for 20 days	Negative with addition of cellulose
Sandstead et al. (1978)	Soft white wheat bran (26 g) or corn bran (26 g) for 28–30 days	Not significantly affected by bran
Guthrie and Robinson (1978)	Wheat bran (14 g) for four weeks	Not significantly affected by bran
Drews et al. (1979)	Cellulose, hemicellulose or pectin (14.2 g) for four days	Significantly lowered by hemicellulose
Papakyrikos et al. (1979)	Cellulose, hemicellulose or wheat bran (10 or 20 g) for seven days	Tendency of all sources to increase fecal zinc loss
Kies et al. (1979)	Hemicellulose (4.2, 14.2, or 24.4 g) for 14 days	Decreased as fiber intake increased; negative with highest level of fiber
Kelsay et al. (1979)	Fruits and vegetables (24 g of neutral-detergent fiber) for 26 days	Negative with fruits and vegetables
Kelsay et al. (1979)	Fruits and vegetables (10, 18, or 25 g of NDF) for 21 days	Significantly lower on 25 g of NDF than on low fiber diet or on 10 g of NDF

a Kelsay (15)

absorbed in animals fed plant protein rather than animal protein (17). This may be attributable to differences in the makeup of the two proteins and/or to the effect due to associated components such as fiber and phytate. In studies with preadolescent girls, Price et al. (18) observed that the absorption of zinc from mixed diet (plant and animal foods) was somewhat higher than that from plant foods (Table X). The results may be somewhat confounded by the differences in dietary protein levels. In a recent study, Sandstrom et al. (14) showed a positive correlation between zinc absorption and the protein content in meals containing milk, cheese, beef and egg in various combinations with wholemeal bread. In another study, Sandstrom et al. (19) compared the effect of hamburger, the bun and the combination of the two on changes in plasma zinc in human subjects. They found that meat alone did not appear to inhibit zinc absorption but the bun (with or without meat) caused a reduction in plasma zinc response. The bun is usually quite low in fiber and phytate but contains about 10% protein.

Effect of certain micronutrients.

The effect of micronutrients on zinc utilization has received little attention. Van Campen (20) demonstrated in vitro that zinc and copper are mutually antagonistic during the absorptive process. Calcium also aggrevates zinc deficiency when added to diets based on plant products that might also contain appreciable amount of phytate. For example, the relative BV of zinc in alkali-treated corn (calcium hydroxide) is reported to be low compared to boiled or raw corn (9). When the diet is free of or low in phytates, zinc absorption in humans consuming high or low calcium levels appears not to be affected (21). Solomons (22) reports a significant reduction in zinc absorption in humans in the presence of nonheme (inorganic) iron (Table XI); heme iron seems to have no such effect. Duncan and Hurley (23) studied the interaction between zinc and vitamin A in pregnant and fetal rats and found that plasma vitamin A in both animals was significantly reduced by low intakes of either zinc or vitamin A. Over the range of doses commonly consumed by man, ascorbic acid was shown (24) to have no demonstratable effect, unlike that on iron, on the absorption of inorganic zinc in man.

Table X. Effect of protein on zinc absorption in adolescent girls[a]

Dietary protein		Zinc		
Source	Amount (g/day)	Intake (mg)	Fecal and urinary losses(mg)	Absorbed (%)
Plant	25	4.83	4.29	11.2
Plant	25	4.53	3.72	17.9
Mixed	46	6.93	4.94	28.7
Mixed	46	6.83	5.09	26.0

[a] Price et al. (18)

Table XI. Zinc and iron (inorganic) interaction in humans[a]

Treatment	Fe:Zn ratio	Plasma Zn (µg/dl)[b]			
		1 hr	2 hr	3 hr	4 hr
a. 25 mg Zn^{++}	0	56+11	85+18	72+16	48+12
b. a + 25 mg Fe^{++}	1:1	18+8	40+9	47+8	36+9
c. a + 50 mg Fe^{++}	2:1	23+2	44+4	38+8	27+8
d. a + 75 mg Fe^{++}	3:1	16+3	27+5	21+4	19+4

[a] Solomons (22)
[b] Rise above fasting plasma levels

Literature Cited

1. Prasad, A.S. "Zinc in Human Nutrition"; CRC Crit. Rev. Clin. Lab. Sci.: Boca Raton, FL., 1977; p. 891
2. Oberleas, D.; Harland, B.F. J. Am. Diet. Assoc. 1981, 79, 433
3. Murphy, E.W.; Willis, B.W.; Watt, B.K. J. Am. Diet. Assoc. 1975, 66, 345
4. Freeland, J.H.; Cousins, R.J. J. Am. Diet. Assoc. 1976, 68, 526
5. Food and Nutrition Board. "Proposed Fortification Policy for Cereal-Grain Products"; National Res. Council/National Acad. Sci.: Washington, DC, 1974; p. 2
6. Ranhotra, G.S.; Loewe, R.J.; Puyat, L.V. Cereal Chem. 1977, 54, 496
7. Marston, R.M.; Peterkin, B.B. Natl. Food Rev. 1980, NFR-9, 21
8. Ranhotra, G.S.; Winterringer, G.L. "The Consumption Pattern of Variety Breads in the U.S. in Variety Breads in the United States (Miller, B. ed.); Am. Assoc. Cereal Chem.: St. Paul, 1981, p. 37
9. Franz, K.B. ; Kennedy, B.M.; Fellers, D.A. J. Nutr. 1980, 110, 2272
10. Ranhotra, G.S.; Lee, C.; Gelroth, J.A. Cereal Chem. 1979, 56, 552
11. Ranhotra, G.S.; Lee, C.; Gelroth, J.A. Nutr. Rep. Intl. 1978, 18, 487
12. Reinhold, J.G.; Parsa, A.; Karimian, N.; Hammick, J.W.: Ismail-Beigi, F. J. Nutr. 1974, 104, 976
13. Morris, E.R.; Ellis, R. J. Nutr. 1980, 110, 2000
14. Sandstrom, B.B.; Arvidsson, A.; Cederbald, A.; Bjorn-Rasmussen, E. Am. J. Clin. Nutr. 1980, 33, 739
15. Kelsay, J.L. Cereal Chem. 1981, 58, 2
16. Anderson, B.M.; Gibson, R.S.; Sabry, J.H. Am. J. Clin. Nutr. 1981, 34, 1042
17. O'Dell, B.L. Am. J. Clin. Nutr. 1969, 22, 1315
18. Price, N.O.; Bunce, G.E.; Engel, R.W. Am. J. Clin. Nutr. 1970, 23, 258

19. Sandstrom, B.; Arvidsson, A.; Bjorn-Rasmussen, E.; Cederbald, A. "Zinc Absorption from Bread Meals in Trace Elements Metabolism in Man and Animals (Kirchgessner, M. ed.) III." Proc. Third Intl. Sym. Arpeitskeirs Fuhr Tierernahrungs Forschung: Weihenstephan, 1978; p. 129
20. Van Campen, D.R. J. Nutr. 1970, 97, 104
21. Spencer, H.; Vankinscott, V.; Lavin, I.; Samachon, J. J. Nutr. 1965, 86, 169
22. Solomons, N.W. J. Am. Diet. Assoc. 1982, 80, 115
23. Duncan, R.J.; Hurley, L.S. J. Nutr. 1978, 108, 1431
24. Solomons, N.W.; Jacob, R.A.; Pineda, O.; Viteri, F.E. Am. J. Clin, Nutr. 1979, 32, 2495

RECEIVED October 5, 1982

Zinc Bioavailability in Infant Formulas and Cereals

B. G. SHAH and B. BELONJE

Banting Research Center, Health and Welfare, Nutrition Research Division, Ottawa, Ontario, Canada, K1A OL2

A slope ratio bioassay based on the total femur zinc in weanling rats fed for 3 weeks the standard ($ZnSO_4$) and the test sources at three levels was developed for the determination of relative biological value (RBV) of zinc. By this assay the RBV of zinc in milk-base formula was 82-86% and in soy-base formula was 53-67%. Although all the conditions of the assay were not met when infant cereals or breakfast cereals were tested, the slopes of the regression lines for the standard and the test sources were indicative of zinc availability: Assay I - $ZnSO_4$, 0.075; mixed infant cereal (wheat, oats and corn), 0.013; breakfast cereal (whole wheat and bran), 0.038; Assay II - $ZnSO_4$, 0.067; infant cereal (barley), 0.0067; (rice), 0.022; (soy), 0.033. The availability of zinc in infant cereals appeared to be less than that of breakfast cereal containing more phytate and neutral detergent fibre. Supplementation of the diet containing mixed infant cereal with 12 mg/kg zinc restored to normal the growth and femur zinc of weanling rats indicating that the subnormal growth and femur zinc observed without supplement, were specifically due to zinc deficiency. In a third assay, the availability of zinc in a breakfast cereal containing rice was greater than cereal containing corn, oats or wheat.

The report of a WHO Expert Committee (1) emphasized the need for information on the bioavailability of zinc, since the recommended intake was dependent on this important factor. Although zinc deficiency in humans was initially reported from Iran and Egypt, adverse effects of marginal or low intakes of zinc by infants and children have subsequently been reported from other parts of the world (2). Apparently the zinc intakes in Iran and Egypt were adequate but the bioavailability was adversely affected by the high level of phytate and fibre in the diet (2). The zinc content of the modern diet is decreased by the use of refined cereal products and the bioavailability of zinc is affected by the addition of plant protein as meat extender or by food additives such as EDTA and polyphosphates. The recent publicity for the beneficial effects of fibre on health has encouraged many people to increase their fibre intake. This may increase the zinc content of the diet but its effect on the availability of zinc in the total diet is uncertain (3). As the number of processed and fabricated foods increases, the need for regulating the composition of foods to ensure nutritional quality increases. Moreover, some cereal products are fortified with zinc in order to increase the zinc intake. There is a need to ensure that the zinc in such products has good availability.

Thus it is evident that for regulatory purposes, a good bioassay for the determination of zinc availability is essential, just as there is an A.O.A.C. hemoglobin repletion test for testing iron source food additives (4).

Development Of The Bioassay

In a preliminary experiment, weanling male Wistar rats were depleted of zinc by feeding a low zinc basal diet (0.6 μg/g zinc) for two weeks and then repleted by adding 12 μg/g zinc as zinc sulphate. The analysis of the zinc content of the different tissues at weekly intervals for four weeks revealed that the body weight and the total femur zinc were the parameters of choice because the responses were linear with duration of feeding. Moreover, the relative errors of the slopes of the regression lines were minimal (5). The results of this experiment also showed that since depletion did not reduce the variability in these parameters, it was not essential for the assay.

The choice of total femur zinc over body weight was made on the basis of linearity between logarithm of total femur zinc and level of dietary zinc from the standard (zinc sulphate) or test source (zinc acetate) over a sufficiently wide range. The response of body weight or body weight gain to dietary zinc from 3 to 12 μg/g was not linear but the logarithm of the total femur zinc at 2 and 3 weeks had a linear relationship with dietary zinc level (Figure 1). At 4 weeks, however, the linearity did not hold. Calculation of the relative biological availability of zinc acetate at 2 and 3 weeks revealed that the 95% fiducial limits were narrower at 3 weeks. Thus a 3 week feeding period was found to be optimal for the bioassay (6).

Application Of The Bioassay To Infant Formulas

The bioassay was used to determine the bioavailability of zinc in milk-base and soy-base infant formulas (7). The results are shown in Figure 2 and Table 1. The low zinc basal diet contained (%): spray dried egg white powder, 20; corn oil, 10; non-nutritive fibre, 3; starch, 25; biotin, 0.0004; vitamin mixture, 1; salt mixture, excluding zinc, 4; dextrose, 37.

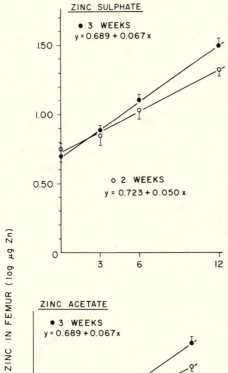

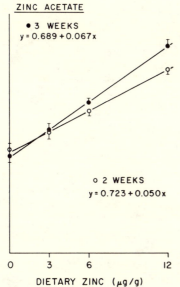

Figure 1.   Regression of log total femur zinc (± S.D.) on dietary levels of zinc at 2 and 3 weeks (6).

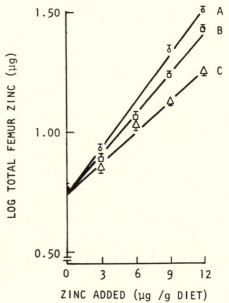

Figure 2.   Log total femur zinc (± S.D.)
of rats fed A, zinc sulfate; B, milk-based;
and C, soy-based infant formula as a
source of zinc (7).

Table I. Relative zinc availability from milk- and
soy-base infant formulas (ZnSO₄ = 100%)

Infant formula base	Relative availability	95% Fiducial limits	
		Lower	Upper
Milk	86	82	91
Soy	67	62	71

Reproduced with permission from Ref. 7. Copyright 1976, American
Institute of Nutrition.

Another milk-base formula had a comparable relative zinc availability of
82% but a second soy-base formula gave a slightly lower value of 53%.
This could be due to a smaller amount of zinc added as zinc sulphate (8).
The lower zinc availability for the soy-base formula than for the milk-
base formula was in agreement with the results obtained by others (9)
using extrinsic labelling with radioactive zinc. This was probably due to
the phytate in the soy isolate (10), although the molar ratio of phytic acid
to zinc was calculated to be no higher than 5. Generally, it has been
accepted that a ratio of 10 or less would not have an adverse effect on
zinc bioavailability (11). Some other factors which might have decreased
the availability of zinc in the soy-base formulas were the higher levels of
calcium (0.6 to 1.4%) and phosphorus (0.5 to 1.0%) than the corresponding
levels (0.33 to 0.45; 0.26 to 0.35%) in the milk-base formulas. Recently,
it has been shown that addition of milk, a source of calcium and
phosphorus to a meal containing soybeans, depressed the absorption of
zinc by men and women (12). This slope ratio bioassay was successfully
employed by others (17) for the determination of the availability of zinc
in soy products.

Application Of The Bioassay To Infant Cereals And Breakfast Cereal

Infant cereals are the first solids usually introduced in an infant's
diet. They provide 12 to 20% of dietary zinc (13). We, therefore, applied
our assay to determine the availability of zinc in some cereals (8,14). The
results are given in Figures 3 and 4. All diets were isonitrogenous and
isocaloric. In Figure 3 are shown the results for one mixed infant cereal
and for one bran breakfast cereal in addition to the two infant formulas
mentioned above. The responses in terms of logarithm of total femur zinc
were linear for all the test foods; but the regression lines did not have a
common intercept. The same observation is true for the results shown in
Figure 4 for three infant cereals containing rice, barley or soy four. Since
one condition required for the assay was not fulfilled the relative
biological availability of zinc in these foods could not be calculated.
However, the slopes of the regression lines, given in Table 2, may be
considered as indicators of zinc availability. From the slopes it is evident
that the bioavailability of zinc in the mixed infant cereal and in the
barley cereal was low, whereas that for the bran breakfast cereal and the

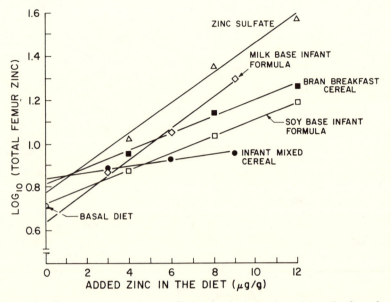

Figure 3. *Log total femur zinc vs. dietary level of zinc from infant foods and a bran breakfast cereal (8).*

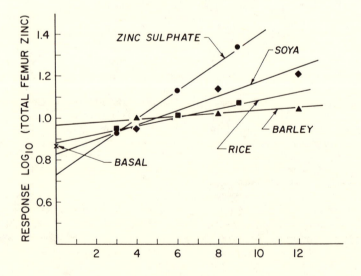

Figure 4. *Log total femur zinc vs. dietary level of zinc from infant cereals (14).*

soy-based infant cereal was much higher. The availability of zinc in the rice infant cereal was intermediate between the above two groups. It is worth noting that the regression lines with smaller slopes had larger intercepts and consequently the requirement of common intercept for the assay was not met.

Table 2. Slopes and intercepts of regression lines for several
infant cereals and one breakfast cereal

Source	Levels, μg Zn/g	Slope of line	Intercept
Assay I			
ZnSO₄	4,8,12	0.075	0.77
Infant Cereal (Wheat, oats and corn)	3,6,9	0.013	0.84
Bran Breakfast Cereal (Whole wheat flour + Bran)	4,8,12	0.038	0.81
Assay II			
ZnSO₄	3,6,9	0.066	0.75
Barley Cereal	4,8,12	0.007	0.97
Soy-flour + Tapioca Starch Cereal	4,8,12	0.033	0.83
Rice Cereal	3,6,9	0.022	0.88

Reproduced with permission from Ref. 14. Copyright 1979, Basel.

Supplementation Of Infant Cereal With Zinc

In order to demonstrate that the low availability of zinc in the mixed infant cereal was the only factor responsible for the reduced food intake and growth of young rats, three groups were fed for three weeks the following diets: (1) Basal diet containing 0.6 μg/g zinc. (2) Diet (1) + mixed infant cereal to provide 9 μg/g zinc. (3) Diet (2) + 12 μg/g zinc from zinc sulphate. All diets were isonitrogenous and isocaloric. The results are summarized in Table 3.

From these results, it is evident that supplementation of the diet containing mixed infant cereal resulted in a body weight gain better than and a femur zinc value comparable to that observed before in weanling rats fed a diet containing 12 μg/g zinc from zinc sulphate (7). The zinc supplementation reduced the phytate/zinc molar ratio of the diet from 10 to 5.

Table 3. Effect of zinc supplementation of a diet containing
mixed infant cereal as a source of zinc

Source and level of zinc added mg/kg		Body weight gain, g	Femur zinc μg
ZnSO$_4$	Cereal		
0	0	14 ± 8[1]	6 ± 1
0	9	65 ± 6	9 ± 1
12	9	141 ± 13	33 ± 3

[1] Mean $\pm$ SD, N = 7

Reproduced with permission from Ref. 14. Copyright 1979, Basel.

Phytate, Fibre And Zinc Bioavailability

It has been reported frequently that fibre and phytate have an
adverse effect on the bioavailability of zinc in foods (3). The cereals in
this study were therefore analysed for phytate (15) and dietary fibre (16).
The results are given in Table 4.

Table 4. Zinc bioavailability and phytate and neutral detergent fibre

Test source (Zn, μg/g)	Ratio of slopes Test/ZnSO$_4$	Phytate P mg/g	Phytate/Zn Molar	ND[1] Fibre %
Infant cereals				
Mixed (24) (Wheat, Corn, Oats)	0.19	0.7	10	2.9
Barley (29)	0.10	1.3	16	6.3
Rice (19)	0.32	0.7	13	1.2
Soy flour + Tapioca starch (30)	0.49	1.2	14	2.4
Breakfast cereal				
Whole wheat + Bran (33)	0.56	2.4	25	10.6

[1] Neutral detergent

These results do not support the concept (10,11) that high phytate or fibre or the molar ratio of phytate to zinc has a consistently adverse effect on zinc bioavailability. If the infant cereals, excluding the one containing soy flour, are considered high phytate, the phytate-zinc molar ratio and high fibre appear to have an adverse effect on zinc availability. The infant cereal containing soy flour, tapioca starch and the breakfast cereal made up of wheat flour and bran show higher zinc availability inspite of high phytate or fibre or both, indicating that there may be intrinsic differences in the binding of zinc with phytate or fibre or both. On the other hand, differences in processing might have caused some of these differences (17).

Almost all the evidence showing that phytate decreases zinc absorption in man and animals is based on pure phytate added to the diet. The effect of natural phytate is variable (18). It has, however, been reported that phytate in bran affected zinc bioavailability in the same way as sodium phytate (19). Dietary fibre in the rural Iranian diet was considered to be the main cause of zinc deficiency in Iran (20). However, the addition of 26 g of fibre from various sources to the American diet did not have any significant effect on the zinc requirements of male adults (21). Similarly, Indian men consuming a diet containing only 10.8 mg of zinc were reported to be in balance in spite of a dietary fibre intake of 50 g per day (22). Moreover, the presence of fibre and phytate in soy flour did not affect the bioavailability of zinc added as zinc carbonate, to the diet of rats (17), although others (23) have reported that the bioavailability of zinc in breakfast cereals depends mainly on their phytate-zinc molar ratio. Our results indicate that there is some, as yet, undetermined difference in the phytate or the fibre of cereals which affects the bioavailability of zinc. It may be some component of dietary fibre (24) or the intrinsic differences in the protein-phytate-mineral complex (10).

Fibre Components And Zinc Bioavailability

To determine if any component of the dietary fibre in the cereals investigated here was correlated with zinc bioavailability, hemicellulose, cellulose and lignin were determined by the method of Mongeau and Brassard (16). The results are summarized in Table 5.

Considering the results of the first three infant cereals, the hemicellulose and cellulose contents appear to be inversely related to zinc availability. Since the lignin content of these foods did not vary appreciably, its effect on zinc availability would not be discernible. Again the infant cereal containing soy flour, tapioca starch and the breakfast cereal consisting of whole wheat flour and wheat bran did not show the adverse effect of high hemicellulose or cellulose or both on zinc availability. Thus, there may be unknown differences in these or other fibre components as they occur naturally, although it has been reported that of cellulose, hemicellulose and pectin, only hemicellulose had an adverse effect on zinc balance in adolescent boys (24) and in adult males (25). In considering the effects of phytate and fibre on the metabolism of zinc or any other mineral nutrient, the role of adaptation, including the presence of microflora and phytase activity must not be neglected (26,27).

Table 5. Fibre components (%) and zinc bioavailability

Test source	Ratio of slopes Test/ZnSO₄	Hemicellulose	Cellulose	Lignin
Infant cereals				
Mixed (Wheat, Corn, Oats)	0.19	1.7	0.5	0.7
Barley	0.10	4.9	0.9	0.5
Rice	0.32	0.6	0.3	0.5
Soy flour, Tapioca starch	0.49	1.0	1.5	0.2
Breakfast cereal				
Whole wheat, bran	0.56	6.6	2.9	1.1

Modification Of The Bioassay

The failure of the zinc bioassay (6) when applied to infant cereals and breakfast cereals (Figures 3 and 4) appeared to be due to the fact that the femur zinc response to poor sources was less than or closer to the lowest level of the standard (ZnSO₄). It is also worth noting that because of the low level of zinc in some cereals, it is not possible to attain even the lowest level of the standard. With a view to investigate the possibility of using low levels of zinc in the bioassay and also to determine the nature of the femur zinc response to a range of dietary zinc levels including those higher than the National Research Council (28) recommendation of 12 µg/g, the following experiment was undertaken.

Weanling male Wistar rats (6 per group) were fed for three weeks diets containing 0, 1, 2, 3, 6, 9, 12, 15, 20, 25 µg per g added zinc as sulphate. The results are shown in Figure 5. From these results, it is evident that the body weight response was linear up to 6 µg/g added zinc, whereas a plot of the same against zinc consumption showed a slight curveture at an intake corresponding to 6 µg/g zinc level. In both plots, the response tends to plateau above 10 µg/g zinc level. These observations were true for body weight gain also and were similar to those reported by others (29). On the other hand, plots of total femur zinc against dietary zinc level or zinc consumption revealed that the latter plot gave better linearity than the former. Both plots showed a decrease in the rate of zinc accumulation in the femur above 15 µg/g level. The difference between the response in terms of body weight and femur zinc was that the former was initially linear and then tended to plateau, whereas the latter was sigmoidal. Because of the nature of the femur zinc response curve, it was necessary to plot the logarithm of the total

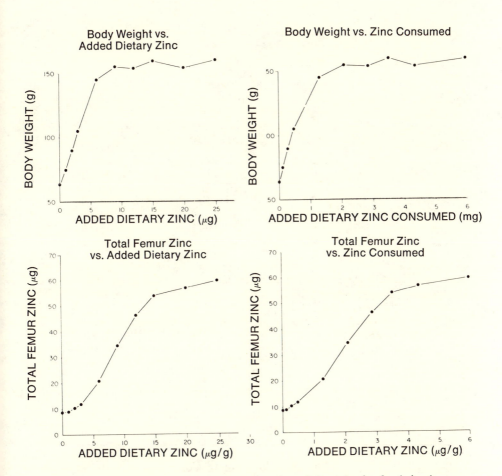

Figure 5. *Body weight and total femur zinc of rats fed varying levels of zinc from zinc sulfate for 3 weeks.*

femur zinc to get linearity (6). Attempts to fit a curvilinear function to such data have been made (29,30) but variable results for relative bioavailability of zinc were obtained using different parameters.

In view of the linearity of response either as body weight or total femur zinc at low zinc levels, it was decided to carry out a zinc bioassay on two breakfast cereals at zinc levels of 2, 4, 6 μg/g. The results are summarized in Tables 6 and 7.

Table 6. Zinc bioassay of breakfast cereals

Source	Added Zn level μg/g	Food consumed in 3 weeks g	Body weight gain g
ZnSO$_4$	0	117+07[a]	22+07
	2	154+25	51+15
	4	184+17	72+12
	6	243+21	106+14
Rice cereal	2	141+10	39+03
	4	177+12	67+05
	6	238+08	108+05
Cereal containing	2	136+07	40+06
oats, corn meal,	4	164+13	61+08
wheat starch	6	222+18	92+06

[a] Mean ± S.D.

Table 7. Relative bioavailability of zinc in breakfast cereals (ZnSO$_4$ = 1.00)

Breakfast cereal	Based on	
	Zinc level	Zinc consumption
Rice	0.97 (0.88, 1.07)[a]	0.99 (0.90, 1.09)
Cereal containing oats, corn meal, wheat starch	0.83[b] (0.74, 0.92)	0.87[b] (0.78, 0.97)

[a] 95% fiducial limits

[b] Significantly less than the estimate for rice cereal (P < 0.05)

The common intercept model was applicable in both cases but in the case of the regression between zinc level and body weight gain, the data for the basal diet could not be included. The breakfast cereal containing corn meal, oats and wheat starch had lower zinc bioavailability than the rice cereal. The neutral detergent fibre content of the former was 3.5% as compared with 1.0% of the latter. The higher fibre content probably had some adverse effect on the availability of zinc. The phytate/zinc molar ratio of a similar breakfast cereal was reported to be only 2 and might not have caused the decrease in zinc availability (23). Since the absorption of zinc by young growing animals is very efficient, a bioassay at low dietary levels will naturally give maximal values for relative bioavailability.

Conclusions

The phytate and neutral detergent fibre in cereal grains and soy products, which make up infant cereals or breakfast cereals, have unknown intrinsic differences which have varying effects on zinc bioavailability. Consequently, the concept of a common phytate/zinc molar ratio below which there is no adverse effect on the bioavailability of zinc cannot be supported from our results.

For a zinc bioassay, the use of low levels of zinc from the standard and the test food sources, appears to eliminate the lack of common intercept, observed with higher levels. This modification of the assay is worth investigating further.

Acknowledgements

The authors appreciate the help of Dr. R. Mongeau, Nutrition Research Division, Health Protection Branch, in the fibre analysis of the cereals. They also thank Mr. S. Malcolm, Food Statistics and Operational Planning, for statistical analysis of the data.

Literature Cited

1.   WHO Expert Committee. "Technical Report Series No. 532. Trace Elements in Human Nutrition"; World Health Organization: Geneva, 1973; p 14.
2.   Prasad, A.S. "Zinc in Human Nutrition"; CRC Press: Boca Raton, Florida, 1979; pp 1-30.
3.   Underwood, E.J. "International Review of Biochemistry, Volume 27, Biochemistry of Nutrition 1A"; Neuberger, A. and Jukes, T.M., Eds.; University Park Press: Baltimore, 1979; pp 207-243.
4.   Association of Official Analytical Chemists. "Official Methods of Analysis of the A.O.A.C."; Washington, D.C., 1980; p 775.
5.   Momcilovic, B.; Belonje, B.; Shah, B.G. Nutr. Rep. Internat. 1975, 11, 445-52.
6.   Momcilovic, B.; Belonje, B.; Giroux, A.; Shah, B.G. Nutr. Rep. Internat. 1975, 12, 197-203.
7.   Momcilovic, B.; Belonje, B.; Giroux, A.; Shah, B.G. J. Nutr. 1976, 106, 913-7.

8.   Momcilovic, B.; Shah, B.G. Nutr. Rep. Internat. 1976, 14, 717-24.
9.   Evans, G.W.; Johnson, P.E. Am. J. Clin. Nutr. 1977, 30, 873-8.
10. O'Dell, B.L.   "Soy Protein and Human Nutrition"; Wilke, H.L.;
     Hopkins, D.T.; Waggle, D.H., Eds; Academic Press: New York, 1979;
     pp 187-204.
11. Oberleas, D.; Harland, B.F. J. Am. Diet. Assoc. 1981, 79, 433-6.
12. Sandström, B.; Cederblad, A. Am. J. Clin. Nutr. 1980, 33, 1778-83.
13. Kirkpatrick, D.C.; Conacher, H.B.S.; Méranger, J.C.; Dabeka, R.;
     Collins, B.; McKenzie, A.D.; Lacroix, G.M.A.; Savary, G. Can. Inst.
     Food Sci. Technol. J. 1980, 13, 154-61.
14. Shah, B.G.; Giroux, A.; Belonje, B. Nutr. Metab. 1979, 23, 286-93.
15. Pons, W.A.; Stansbury, M.F.; Hoffpanir, C.L.   J. Am. Off. Anal.
     Chem. 1953, 36, 492-504.
16. Mongeau, R.; Brassard, R. J. Food Sci. 1982, 47, in press.
17. Erdman, J.W. J.A.O.C.S. 1979, 56, 736-41.
18. Shah, B.G.   "Symposia from the XII International Congress of
     Nutrition"; Alan, R. Liss, Inc.: New York, 1981; p 199-208.
19. Morris, E.R.; Ellis, R. J. Nutr. 1980, 110, 2000-10.
20. Reinhold, J.G. "Proc. 9th Int. Congr. Nutr., Mexico, 1972"; Karger:
     Basel, 1975; p 115.
21. Sanstead, H.H.; Kleavay, L.M.; Jacob, R.A.; Munoz, J.M.; Logan,
     Jr., G.M.; Reck, S.J.; Dintzis, F.R.; Inglett, G.E.; Shuey, W.C.
     "Dietary Fibres: Chemistry and Nutrition"; Inglett, G.E.; Folkehag,
     S.I., Eds.; Academic Press: New York, 1979; p 147.
22. Rao, C.N.; Rao, B.S.N. Nutr. Metab. 1980, 24, 244-54.
23. Morris, E.R.; Ellis, R. Cereal Chem. 1981, 58, 363-6.
24. Drews, L.M.; Kies, C.; Fox, H.M. Am. J. Clin. Nutr. 1979, 32, 1893-
     7.
25. Kies, C.; Fox, H.M.; Beshgetoor, D. Cereal Chem. 1979, 56, 133-6.
26. King, J.C.; Stein, T.; Doyle, M. Am. J. Clin. Nutr. 1981, 34, 1049-
     55.
27. Anderson, B.M.; Gibson, R.S.; Sabry, J.H. Am. J. Clin. Nutr. 1981,
     34, 1042-8.
28. National Research Council. "Nutrient Requirements of Laboratory
     Animals, Report No. 10"; National Academy of Sciences:
     Washington, D.C., 1978; p 23.
29. Franz, K.B.; Kennedy, B.M.; Fellers, D.A. J. Nutr. 1980, 110, 2263-
     71.
30. Franz, K.B.; Kennedy, B.M.; Fellers, D.A. J. Nutr. 1980, 110, 2272-
     83.

RECEIVED August 23, 1982

Zinc Absorption from Composite Meals

WENCHE FRØLICH
Norwegian Cereal Institute, National Institute of Technology, Oslo 1, Norway

BRITTMARIE SANDSTRØM
Sahlgrenska Sjukhuset, Institutt for klinisk näringslära, 41345 Gøteborg, Sweden

Zinc absorption was measured from meals based on
wheat bread. A lower absolute amount of zinc was
absorbed from a low zinc white bread (72% extraction
rate) than from a whole-meal bread with a higher
zinc content (100% extraction rate). When the two
types of bread where enriched with zinc chloride,
more zinc was absorbed from the white bread. By increasing the zinc content of the whole meal
bread in form of animal protein sources, such as
milk and cheese, the absorption increased to the
same level as that from zinc-enriched white bread.
Addition of calcium in the form of milk products improved the absorption of zinc from a meal with whole
meal bread.

Zinc is well established as an essential nutrient and signs of
zinc deficiency in man has been reported (1).
 In Norway zinc deficiency seems to be unusual, although a
suboptimal supply in certain groups in the population might occur.
The daily dietary requirement of zinc depends not only on the
physiological requirements of zinc, but also on the composition of
the meals.
 In Norway there is essentially no fortification of foodstuffs
(exception iron in goat cheese, vitamin A and vitamin D in
margarine, and iodine in salt).
 We have in Norway an official nutrition and food policy since
1975 (2). In this Whitebook from the Government to the Parliament, the authorities are trying to stimulate the consumer and
producer towards a diet which should be more suitable.
 The nutrition and food policy should coordinate several important objectives and considerations. Members of the National
Nutrition Council which are appointed from the Social and Health
Department are working with guidelines of this policy.
 The best way to reach this nutritional goal is to give the
consumer information and education about which foods are the best

0097-6156/83/0210-0211$06.00/0

sources for the different nutrients, and to encourage them to eat
more of these products, which usually means more unrefined products in the case of minerals.

The success in reaching the goals which are predicted in the
nutrition policy, depend on the research in progress. An increased knowledge of the relation between nutrition and health
requires comprehension of the nutrients, among others how an insufficient supply of different nutrients could lead to or be a
part of the development of diseases.

Even if the diet in Norway must be considered nutritionally
adequate, generally covering the required essential nutrients,
there could in certain groups in the population exist an insufficient supply of certain minerals. An increased knowledge
about the different minerals in our diet is important, and a lot
of work is done in many areas, e.g. on iron, zinc, selenium and
chromium.

Improving the diet is in itself not enough. How an optimal
bioavailability of the minerals could be obtain is as important
as knowledge about the content of these minerals.

This paper will deal with the mineral zinc, and how a meal
could be composed in such a way that a maximal absorption could
be the result.

The amount of zinc that the diet must contain, depends on the
availability of zinc for absorption. The WHO committee gives no
data for diets, and the suggested daily dietary requirement range
from 2.8 mg in infancy to 54.5 mg for lactating women. The
United States Food and Nutrition Board suggest a daily intake of
15 mg a day for adults (3).

The average intake of zinc in Norway is 14.2 mg per person
per day, where 30% is coming from cereals. This is except for
meat, one of the largest single sources of zinc in our diet.

An increase in consumption of whole grain flour products is
one of the nutritional aims in Norway. The high content of dietary fiber or factors associated with it, however, present in bran
and whole grain flour, may interfere with the bioavailability of
minerals as suggested by several authors (4, 5, 6, 7, 8).
Especially phytic acid has been claimed to be a potent inhibitor
of zinc absorption and zinc deficiency on high intake of phytic
acid have been shown in different animals (9, 10).

Knowledge about how an optimal absorption could be obtained
from whole grain products is of great value, since this is an important source of zinc in our diet. It is important to notice
that it is not only the bioavailability of zinc from bread which
is important, it is the bioavailability of zinc from the composite
meal where bread is one of the components which gives the most
interesting and useful information.

In Sweden and Norway the eating pattern are rather common,
and we have more or less the same aim for our nutrition policy.
More unrefined cereal products is an important factor in this
effort for better eating habits.

The experiments which are to be described in this paper, were carried out by Dr. Brittmarie Sandstrøm from the Department of Clinical Nutrition at the University of Gothenburg, Sweden, (11).

The aim of the study
The aim of the present study was

1. To study the effect of phytic acid, calcium and protein in bread meals in various combination, on the zinc absorption in man.
2. To study the effect on zinc absorption after enrichment with zinc chloride of bread with different extraction rate.

As zinc deficiency has been observed on high intakes of whole flour bread, 50% of the total energy of the meal was covered by bread.

Materials and methods

Subjects
Healthy medical or dietetic students and laboratory personal, all with normal serum zinc levels, volunteered for the study.
Thirthy-five women and thirty-one men between nineteen and sixtyone years of age (average 22 years) participated in the experiment. No apparent differences in zinc intake were found and they were not consuming mineral supplements.

Absorption measurment
The meals were extrinsically labelled by added 65 Zn. The rationale for this method is that a complete isotope exchange takes place between the added radioactive zinc isotope and the zinc present in the meal. Measurements of the uptake of radioactive iron isotopes in blood or in the whole body have been used for many years in studies of iron absorption. (12, 13, 14). The absorption in the present study is determined from measurement of the whole body retention of the radioisotope. However, this can not be done until the non-absorbed fraction of the isotope has left the body. During this periode of time some of the initially absorbed ^{65}Zn has been extrected. A correction of retention data for this endogenous zinc excretion is made, based on figures for the rate of excretion after intravenously injected ^{65}Zn.
In healthy subjects the differences in the rate of zinc turnover are relatively small and a standard correction can be made. Otherwise repeated measurements of the rate of excretion can be used to extrapolate retention to absorption values at the time for intake of the meal. The retention measurements is done after two weeks to allow for long bowel transit time. The endogenous extretion of 65 Zn during this period is only about 10%. The radionuctide technique with 65 Zn and whole body counting are described in more detail in an other paper (15).

Preparation and composition of the meals

In the present study, bread in the form of rolls was prepared from whole grain flour (100% extraction rate) with a zinc content of 22 mg/kg and from white wheat flour of about 72% extraction rate, where the zinc content was 7 mg/kg. The water added to the dough, contained an amount of almost carrier free isotope solution, corresponding to 0.25 μCi of [65] Zn for each roll. The isotope solution was made by diluting a stock solution of 65 Zn Cl$_2$ in 0.1 M HCl (0.1 to 0.5 Ci/g Zn, Radiochemical centre, Amersham, England) with physiological saline to a final radioactivity of 1 μ Ci/ml.

Each test meal contained two rolls and 200 ml liquid. All meals except for 2 meals were prepared to have a total zinc content of 3.5 mg. In those meals which did not have a natural content of 3.5 mg zinc, zinc chloride, was added during preparation of the bread. The energy contents in the different meals were equalized to 380 kcal by serving various amounts of butter with the bread and various liquids (3% milk, 0.5% milk or water).

The calcium content of the meals was increased by adding milk or lowfat cheese. In the studies on the effect of protein, the animal protein sources were milk, cheese, boiled beef and boiled egg. To further increase, the protein content without at the same time increasing the calcium or zinc content, sodium caseinate was used in one meal.

Experimental design

On the basis on the isotope technique described (15), the following experimental design was applied:

1. The meals were prepared in advance, extrinsically labelled and measured in the whole body counter.
2. Each subject fasted for 12 hours.
3. The test meal was eaten, and no other food was allowed for the following three hours.
4. Retention was measured after 14 days and the figures were corrected for endogenous excretion.
5. Absorption data were expressed in per cent and absolute amount absorbed.

Justification of comparison of per cent absorption figures can only be done if the zinc contents of the meals are identical.

When meals of different zinc contents are compared it is necessary to express the absorption in terms of absolute amounts of zinc absorbed.

Chemical methods

The total zinc content was determined by atomic absorption spectrophotometry (Perkin Elmer Model 360), after dry-ashing at 500°C. The calcium analysis was also done by atomic absorption spectrophotometry, but after wet-ashing in sulphuric acid and hydrogen peroxide.

Nitrogen analysis was performed by a micro-Kjeldahl technique in a Technicon Autoanalyser.

Phytic acid determinations were performed by a modification of the procedure described by Holt (16, 17).

The radioactivity in each individual portion of bread was measured in the whole-body counter before serving.

TABLE I

Composition of the meals.

Meal no and composition	Zinc content (mg)		Zn absorption	
	Total	Added	%	mg
1. White bread	0.4	–	38.2[a] (29.9-52.0)[b]	0.15 (0.12-0.21)
2. White bread	3.6	3.1	13.2 (8.7-24.2)	0.48 (0.31-0.87
3. White bread, milk, cheese	3.3	1.1	15.7 (6.9-25.3)	0.52 (0.23-0.83)
4. Whole flour bread	1.3	–	16.6 (19.1-23.2)	0.22 (0.12-0.30)
5. Whole flour bread	3.5	2.2	8.2 (5.7-11.3)	0.29 (0.20-0.40)
6. Whole flour bread milk	3.1	1.0	9.9 (5.6-14.4)	0.31 (0.17-0.45)
7. Whole flour bread beef	3.2	0.8	10.6 (6.0-15.6)	0.34 (0.19-0.50)
8. Whole flour bread cheese, milk	3.2	–	14.0 (8.7-21.8)	0.45 (0.28-0.70)
9. Whole flour bread egg, milk	3.5	0.4	14.8 (10.0-22.9)	0.52 (0.35-0.80)
10. Whole flour bread cheese,milk,casein	3.5	–	16.8 (10.7-26.3)	0.59 (0.37-0.92)

[a]Mean   [b]Range

Results and discussion
    The composition of 10 different meals are shown in Table I.
This table also includes the zinc content, the per cent absorption and the absolute amount absorbed.
    As can be seen, there is a significantly higher per cent
absorption in meals consisting of white bread (low extraction
rate) than in meals consisting of whole flour bread, 38.2%
versus 16.6%. The amount of zinc was however considerably higher
in whole flour bread (1.3 mg versus 0.4 mg), which causes a
higher absolute absorption of zinc from whole flour bread.
(0.22 mg versus 0.15 mg). When these two types of bread were
enriched with zinc chloride, resulting in a zinc content of
3.6 mg and 3.5 mg respectively, an increased absorption was found
in both of breads. The absorption however was higher from the
enriched white bread than from the enriched whole flour bread.
(0.48 mg (13.2%) versus 0.29 mg (8.2%)).
    Consequently if the white bread is enriched with zinc this
could be a better zinc source than whole flour bread enriched to
the same level.

TABLE II

Zn-absorption from meals with different protein content

Meal no and composition	Zn content (mg) Total	Added	Protein content (g)	Zn absorption %	mg
2. White bread	3.6	3.1	5.2	13.2 (8.7–24.2)	0.48 (0.31–0.87)
5. White flour bread	3.5	2.2	6.0	8.2 (5.7–11.3)	0.29 (0.20–0.40)
8. Whole flour bread	3.2	–	19.3	14.0 (8.7–21.8)	0.45 (0.28–0.70)
10. Whole flour bread cheese, milk,casein	3.5	–	26.7	16.8 (10.7–26.3)	0.59 (0.37–0.92)

    When adding animal protein in the form of milk, egg and
cheese to the whole flour meal, the zinc absorption was increased.
(Table II). It seems like the binding of zinc to phytic acid and
fibercomponents is overcome if enough animal protein is served
together with the whole flour bread.
    The protein content of whole flour bread meal with 3.5 mg
zinc is 6 g and the absorption is 0.29 mg (8.2%). By increasing
the protein content to 19.3 g cheese and milk and keeping the same
zinc content, the absorption increased to 0.45 mg (14%).

By increasing the protein content further to 26.7 g with the zinc concentration still at 3.5 mg, the absorbed amount of zinc was 0.59 mg (16.8%).

TABLE III

Zn absorption from meals with different Ca content

Meal no and composition	Zn content (mg) Total	Added	Protein gr	Calcium mg	Zn absorption %	mg
5. Whole flour bread	3.5	2.2	6.0	19	8.2 (5.7-11.3)	0.29 (0.20-0.40)
6. Whole flour bread, milk	3.1	1.0	12.6	292	9.9 (5.6-14.4)	0.31 (0.17-0.45)
7. Whole flour bread, beef	3.2	0.8	12.6	23	10.6 (6.0-15.6)	0.34 (0.19-0.50)

In animal experiments it has been shown that a high calcium content in combination with the phytic acid in whole flour bread will decrease the zinc absorption (10). When adding milk and milk products to the whole flour bread in Dr. Sandstrøms experiments the same positive effect was seen from the protein despite of the higher calcium content (Table III).

Meal no. 6 and no. 7 are showing this effect, the protein content is in both meals 12.6 g, while the Ca content in no 6 is about 10 times as high as in no. 7. The zinc absorption however

TABLE IV

Zn absorption from white bread and different protein content

Meal no and composition	Zn content (mg) Total	Added	Protein gr	Zn absorption %	mg
2. White bread	3.6	3.1	5.2	13.2 (8.7-24.2)	0.48 (0.31-0.87)
3. White bread cheese, milk	3.3	1.1	18.6	15.7 (6.9-25.3)	0.52 (0.23-0.83)

is almost the same 9.9% and 10.6%, which in both case are higher than without protein (8.2%).

No increase in zinc absorption was seen when protein was added to the white bread meals. (Table IV).

It is possible that the protein competes with the binding to phytic acid or fiber components in whole flour bread rather than facilitating zinc absorption.

In conclusion, the binding of the zinc due to whole flour bread is not a serious problem in a diet with a low bread and a high protein content such as in normal diets of people in industrialized countries.

By increasing the fermentation period when making the dough for the bread, the decreased zinc absorption in whole flour bread can be overcome. The phytic acid is in this way broken down, and the zinc better available absorption.

In Table V is shown the Zn absorption of the meals after different fermentation on periods.

The meals were breakfast meals, composed of sour milk, white bread and bran or bread with bran in it.

TABLE V

Zn absorption from bread with different
fermentation period

	Zn (mg)	Phytic acid mol	Zn absorption % $(\overline{X}-SD)$	mg
Sour milk, white bread	1.9	500	$9.6^+1.5$	$0.18^+0.03$
Sour milk bread w/ bran yeast 45 min.	1.8	420	$11.9^+2.5$	$0.21^+0.04$
Sour milk bread w/ bran yeast 15 h + 45 min.	1.8	100	$19.8^+4.9$	$0.35^+0.08$

This could be of importance in countries where whole flour bread is a main component in the diet and the supply of animal protein is limited.

Conclusion

Most of our knowledge about factors affect zinc absorption (9, 10, 18) comes from studies on animals. The most extensive studies are made on the affect of phytic acid, which is present in vegetables and unrefined cereal products. Decreased zinc absorption has been found both when substantial amounts of sodium-phytate are added to the diet and when the diets are based on phytic-containing food, such as soy beans. The explanation of this is that a complex between Zn and phytic acid is made, resulting in a decreased bioavailability.

In human studies done with leavened and unleavened whole grain bread (7, 19), the conclusions drawn were the same to those drawn for animal studies. The importance of phytic acid was later questioned by the same authors and decreased absorption was described to the fiber content in the diets (20). A negative zinc balance was also acribed to the fiber content when a diet containing fruits and vegetables (21). Contrary to this is a study where zinc balance was not affected even with a high content of dietary fiber (22), where animal protein also showed a positive correlation.

Even if there is a lot of unanswered questions around bioavailability of zinc from the diet, these studies could give us important information for obtaining an optimal zinc absorption from composite bread meals.

First of all these studies do not support a change from whole flour bread to a more refined bread with a lower zinc content. Even if there is a lower per cent absorption of zinc from whole flour bread, the absolute amount of zinc absorbed from the unrefined products is higher, because of the considerable higher content of zinc in these products.

When the two types of bread were enriched to the same zinc level the absorption was higher from the white bread. If the zinc content of a meal based on whole flour bread was increased in the form of animal protein as milk and milk products, the absorption increased to the same level as that from zinc enriched white bread. Consequently the conclusion drawn from this study shows that in a mixed balanced meal the zinc-binding of whole flour bread has negligable nutritional significance.

Four factors are of importance for the absorption and utilization of a mineral from the diet in humans:

1. The concentration of the mineral in the diet
2. The concentration of other minerals and nutrients in the diet
3. The bioavailability of the mineral in the diet.
4. The individual person's need for the mineral.

There is a lot of unanswered and important questions around bioavailability. In the literature there is a tendency to describe upper limits for availability of the minerals in the diet. Do the "bioavailabilities" represent reliable measurements of the amount of these elements available for absorption, or do they only give values for the average absorption for the particular circumstances of the studies? Is it possible to describe upper limits for availabilities of minerals? Do we really increase the absorbed and utilazed amount of a mineral by fortification with the mineral in question? Do we forget that there is a cooperation between all the elements present in the diet which determine the bioavailability?

An other important factor is that the body might regulate the absorption of the mineral after the body's own need.

The significance of different inhibitors in the diet are the most common way to study the bioavailability of a mineral. The inhibitors of greatest consequence in our diet is as mentioned earlier fiber and phytic acid.

As important as studying these inhibitors significance in a mineral deficiency situation it might be to determine these components value as protecting factors in toxic conditions. Might be the body's own regulation system is put out of function when a mineral is given in pure form without at the same time give an increased amount of other minerals and nutrients. This is one of the reasons why Norway has a very restrictive fortification policy of foods. We think we can have a better control over the intake of minerals without fortifying, rather than give dietary advice to vulnerable groups in the population.

Studies of zinc absorption from single meals can be used to identify dietary factors influencing the degree of absorption. They can, however, only give suggestions for the importance of these factors for the whole diet. Furthermore it is uncompletely known how the absorption of zinc is regulated. Will the body compensate a low absorption from one meal with a higher absorption from the next? The only way to overcome this problem is long term balance studies.

To obtain complete zinc balance it is important to include all components which might give losses. Losses from the skin are seldom included in these types of studies, and as a consequence it is difficult to talk about true balance. What is determined is a relative balance. Most of the zinc balance studies published so far, are short term experiments, and it is therefore difficult to draw any conclusions about the value of different diets as complete zinc balance is probably not obtained.

From the data obtained for zinc absorption in the studies presented here, the contents of zinc, protein and phytic acid were useful in estimating the degree of zinc absorption. The main determinant, however, seems to be the total zinc content.

For practical use for the consumer could be concluded: The binding of zinc due to whole-meal bread is not a serious problem in a well-balanced diet with a relative low bread and a high protein content such as in the normal diets of people living in industrialized nations.

Literature Cited

1. Prasad, A. S. "Trace elements in human health and disease. I. Zinc and Copper"edited by A. S. Pasad. New York: Academic Press 1976, p. 1.
2. Anonymous, 1975. "On Norwegian nutrition and food policy". Whitebook to the Parliament, november 1975. The Royal Ministry of Agriculture, Oslo.
3. National Research Council. Food and Nutrition Board.

Recommended Dietary Allowances, 9th Ed., National Academy
of Sciences, Washington, D.C. 1980.

4. Maleki, M. Iran. Ecol. Food Nutr. 1973, 2, 39.

5. Prasad, A. S.; Miale Jr. A.; Farid, Z.; Sandstead, H. H.;
    Schulert, A. R. J. Lab. Clin. Med. 1963, 61, 537.

6. Prasad, A. S.; Schulert A. R.; Niale Jr. A.; Farid Z.;
    Sandstead, H. H. Am. J. Clin. Nutr. 1963, 12, 437.

7. Reinhold, J. G.; Hedayati, H.; Lahimgarzadeh, A.; Nast, K.
    Ecol. Food Nutr. 1973, 2, 157.

8. Ronaghy, H. A.; Moe, P. G.; Halsted, J. A. Am. J. Clin. Nutr.
    1968, 21, 709.

9. Davies, N. T.; Olpin, S. E. Br. J. Nutr. 1979, 41, 591.

10. O'Dell, B. L. Am. J. Clin. Nutr. 1969, 22, 1315.

11. Sandstrøm, B.; Arvidson, B.; Cederblad, Å.; Bjørn-Rasmussen,
    E. Am. J. Clin. Nutr. 1980, 33, 739.

12. Cook, J. D.; Layrisse, M.; Martinez-Torres, C.; Walker, R.;
    Monsen, E.; Finch, C. A. J. Clin. Invest. 1972, 51, 805.

13. Hallberg, L.; Bjørn-Rasmussen, E. Scand. J. Heamatol. 1972,
    9, 193.

14. Layrisse, M.; Martinez-Torres, C. Am. J. Clin. Nutr. 1972,
    25, 401.

15. Arvidson, B.; Cederblad, Å.; Bjørn-Rasmussen, E.; Sandstrøm,
    B. Internat. J. Nucl. Med. Biol. 1978, 5, 104.

16. Holt, R. J. Sci. Food Agric. 1955, 6, 136.

17. Davies, N. T.; Reid, H. Br. J. Nutr. 1979, 41, 579.

18. Davies, N. T.; Hristic, V.; Flett, A. A. Nutr. Dep. Int.
    1977, 15, 207.

19. Reinhold, J. G.; Nasr, K.; Lahimgarzadele, A.; Hedayati, H.
    Lancet. 1973, 1, 283.

20. Reinhold, J. G.; Ismail-Beigi, F.; Faradji, B. Nutr. Dep.
    Int. 1975, 12, 75.

21. Kelsay, J. L.; Jacob, R. A.; Prather, E. S. Am. J. Clin.
    1979, 32, 2307.

22. Sandstead, H. H.; Munoz, J. M.; Jacob, R. A.; Klevay, L. M.;
    Reck, S. J.; Logan, Jr., G. M.; Dintzis, F. R.; Inglett,
    G. E.; Shuey, W. C. Am. J. Clin. Nutr. 1978, 31, 1980.

RECEIVED October 13, 1982

Zinc Balances in Humans During Different Intakes of Calcium and Phosphorus

HERTA SPENCER, LOIS KRAMER, and DACE OSIS

Veterans Administration Hospital, Metabolic Section, Hines, IL 60141

Animal studies indicate that calcium and phosphorus inhibit the availability of zinc for absorption. In the present study the effect of calcium and phosphorus on zinc metabolism was investigated in adult men by determining metabolic balances of zinc during different intakes of calcium and phosphorus. Three intake levels of calcium, ranging from 200 to 2000 mg/day, and two intake levels of phosphorus (800 and 2000 mg/day) were used during a constant dietary zinc intake of 14.5 mg/day. Increasing the calcium intake from 200 to 2000 mg and increasing the phosphorus intake from 800 to 2000 mg/day had no effect on urinary or fecal zinc excretion nor on zinc retention. ^{65}Zn absorption studies confirmed these results. The observations made in man are in contrast with results obtained in animal studies.

Interaction between zinc and calcium has been demonstrated in several animal studies (1-6). It has been shown that calcium antagonizes the biological effects of zinc and that calcium reduces the availability of zinc for absorption. This decrease in zinc absorption resulted in severe malnutrition and parakeratosis (1-7). Several studies have conclusively shown that the calcium-zinc antagonism studied in animals is due to excess phytic acid in the diet (8, 9). However, in the absence of phytic acid, excess dietary calcium per se has also been shown to decrease the intestinal absorption and the retention of zinc in rats (10). This inhibitory effect on zinc absorption, induced by calcium, was further enhanced by the addition of phosphorus to the high calcium intake (11).

Little information is available in man on the effect of either calcium or of phosphorus on zinc metabolism and specifically little is known on the effect of these minerals on the intestinal absorption of zinc in humans. In a study carried out by

0097-6156/83/0210-0223$06.00/0

this group, using ^{65}Zn as the tracer, a high calcium intake of 2000 mg calcium per day did not decrease the intestinal absorption of zinc in man compared with the absorption during a calcium intake of 200 mg per day (12). However, metabolic balances of stable zinc were not determined at that time. In a study carried out by others, using dairy products with the breakfast meal, namely, milk and cheese, both of which contain calcium and phosphorus, a decrease of the intestinal absorption of zinc was observed which was estimated by zinc plasma levels; however, zinc absorption studies were not carried out (13). This same study has also demonstrated that phosphate per se had a similar effect in decreasing the intestinal absorption of zinc. It should be mentioned that this same study also reported that other substances, such as coffee, decrease the intestinal absorption of zinc in man.

In the present study complete metabolic balances of zinc, calcium, and phosphorus were determined in adult males under strictly controlled dietary conditions in the Metabolic Research Ward in order to determine the effect of these minerals, which were used singly and combined, on the availability of zinc for absorption. The metabolic diet and the daily fluid intake were kept constant and the diet was analyzed for zinc, calcium, and phosphorus throughout all study phases. The basal diet contained an average of 14.5 mg zinc, 850 mg phosphorus, and 230 mg calcium. The calcium intake was adjusted to appropriate levels by the addition of calcium gluconate tablets to the constant diet. Complete collections of urine and stool were obtained for several weeks from the start of the study. Metabolic balances of zinc, calcium, and phosphorus were determined in each 6-day metabolic study period by analyzing the diet and the excreta for these elements. Calcium and zinc were analyzed by atomic absorption spectroscopy (14, 15) and phosphorus by a modification of the method of Fiske and SubbaRow (16). The net or apparent absorption of zinc during the different intakes of phosphorus and calcium was determined by relating the fecal zinc excretion to the zinc intake according to the following formula:

$$\text{Net absorption, } \% = \frac{\text{Zinc Intake} - \text{Fecal Zinc}}{\text{Zinc Intake}} \times 100$$

Several zinc absorption studies, using oral doses of ^{65}ZnCl$_2$ were carried out. ^{65}Zn plasma levels were determined serially on the day of the oral administration of the ^{65}Zn tracer. Urinary and fecal ^{65}Zn excretions were determined for approximately 15 days. The subjects studied were fully ambulatory males who were in good nutritional state. They were normal according to all clinical and laboratory criteria, including the serum levels of zinc, calcium, and phosphorus. The effect of three intake levels of calcium on the zinc balance was studied, namely, of 200, 900,

and 2000 mg calcium per day. In studies of the effect of phosphate on zinc metabolism the increased intake of phosphorus was
due to the addition of sodium glycerophosphate to the constant
diet. The high phosphorus intake was 2000 mg per day compared
to a dietary intake of 800 mg per day in the control study.
During the addition of either calcium or phosphorus all other
dietary constituents remained unchanged. The duration of each
study phase was several weeks both in control studies and during
the higher intakes of calcium or phosphorus.

Effect of Calcum on the Zinc Balance

    Table I shows data of the effect of different intake levels
of calcium on the zinc balance in three patients. During a low
calcium intake of approximately 200 mg per day the major pathway
of zinc excretion is via the intestine, the urinary zinc excretion is low ranging from 0.7 to 0.9 mg per day. Increasing the
calcium intake from 200 to 900 mg per day did not result in any
appreciable changes of the fecal zinc excretion. These excretions reflected the slightly lower zinc intake during the higher
calcium intake. There was no change of the urinary zinc excretion nor of the zinc balance (Patient 1). This lack of change of
the fecal zinc excretion was observed in the presence of a marked
increase of the fecal calcium excretion. The lack of change of
the urinary zinc excretion was observed in the presence of an increase in urinary calcium excretion. When the calcium intake was
increased further, from 200 mg to 2000 mg per day (Patient 2),
there was again no change of the fecal or urinary zinc excretion
nor of the zinc balance. During this high calcium intake of
2000 mg calcium per day (Patient 2), the fecal calcium excretion
was even higher than during the 900 mg calcium intake and the
presence of these large amounts of calcium in the intestine did
not affect the fecal zinc excretion. In the third study carried
out in a patient with idiopathic hypercalciuria (urinary calcium
370 mg per day during a calcium intake of approximately 800 mg
per day), the increase of the calcium intake from 800 to 2000 mg
per day again did not result in any change of the urinary or
fecal zinc excretions nor of the zinc balance, despite the fact
that the urinary and fecal calcium excretions increased, particularly the latter. The net absorption of zinc during the different calcium intakes varied and these values ranged from -1%
of the intake during a 2000 mg calcium intake to 16% on the same
calcium intake. The phosphorus intake during all three studies
was constant, 800 mg/day.
    These studies, as well as additional studies carried out in
this Research Unit, have shown that increasing the calcium intake
up to 2000 mg per day does not affect the excretions or retention
of zinc in man. These observations are in contrast to results
obtained in animal studies (10) which have shown that a high calcium intake per se, in the absence of phytic acid, can decrease

TABLE I

EFFECT OF DIFFERENT INTAKE LEVELS OF CALCIUM ON THE ZINC BALANCE

PATIENT AND STUDY	STUDY DAYS	ZINC, mg/day				Zn, NET ABSORPTION %	CALCIUM, mg/day			
		Intake	Urine	Stool	Balance		Intake	Urine	Stool	Balance
Patient 1										
Low Calcium	60	13.8	0.7	13.2	−0.1	4	198	102	220	−124
Normal Calcium	42	11.9	0.5	11.8	−0.4	1	947	149	637	+161
Patient 2										
Low Calcium	36	17.2	0.9	16.3	0	5	212	57	197	− 42
High Calcium	42	16.0	0.7	16.1	−0.8	−1	2022	176	1772	+ 74
Patient 3										
Normal Calcium	48	15.3	0.9	13.6	+0.8	11	838	370	562	− 94
High Calcium	48	14.6	0.9	12.2	+1.5	16	1903	407	1209	+287

Phosphorus intake = 800 mg/day in all studies

the intestinal absorption of zinc. It should be stated that the type of diet used in our studies had a low phytate content. The difference in results obtained in animals and man is most likely due to the fact that the amount of calcium used in animal studies, relative to the total body weight, was considerably higher than the calcium intake levels used in our studies in humans. It should also be pointed out that in terms of a high calcium intake in man it would be most unlikely that the human diet would contain more than 2000 mg calcium per day. No data are available, to our knowledge, on the effect of calcium intakes greater than 2000 mg per day on the zinc balance in man.

Effect of Phosphorus on the Zinc Balance

Table II shows data of the effect of a high phosphorus intake on the zinc balance. The phosphorus supplements were given to three patients during different calcium intakes, namely, during a low calcium intake of 200 mg per day and during higher calcium intakes of 800 mg and 2000 mg calcium per day. The phosphorus intake of the subjects studied was approximately 900 mg per day in the control study and was 2000 mg per day during the high phosphorus intake. In the control study, during a low calcium intake and a normal phosphorus intake of 900 mg per day and a dietary zinc intake of 17 mg per day, the urinary zinc excretion was relatively high, 1.6 mg/day, the fecal zinc excretion was in the expected range and the zinc balance was positive, +1.6 mg/day. In the experimental study, when the phosphorus intake was increased by approximately 1000 mg per day, there was little change of the urinary or fecal zinc excretions and, therefore, the zinc balance also did not change. In the study carried out during a calcium intake of 800 mg per day and a dietary zinc intake of 13.6 mg per day, the urinary zinc excretion was normal, the fecal zinc excretion reflected the zinc intake, and the zinc balance was slightly negative, -0.7 mg per day. With the addition of 1000 mg of phosphorus the urinary zinc excretion remained unchanged, the fecal zinc excretion decreased slightly, and the zinc balance became slightly positive during this high phosphorus intake. In the study carried out during a high calcium intake of 2000 mg per day and a high phosphorus intake of 2000 mg per day, and a dietary zinc intake of 14.6 mg per day, the following observation was made. In the high calcium control study the urinary zinc excretion was slightly higher than normal, 0.9 mg per day, the fecal zinc excretion reflected the dietary zinc intake and the zinc balance was positive, +1.5 mg per day. During the addition of 1000 mg phosphorus to the high calcium intake of 2000 mg per day, there was no change of the urinary zinc excretion and the fecal zinc excretion increased slightly; however, the dietary zinc intake was also somewhat higher than in the control study. The zinc balance remained unchanged during the high phosphorus intake which was given during a high calcium

TABLE II

EFFECT OF DIFFERENT INTAKE LEVELS OF PHOSPHORUS ON THE ZINC BALANCE

PATIENT	CALCIUM INTAKE mg/day	STUDY DAYS	ZINC, mg/day				Zn, NET ABSORPTION	PHOSPHORUS, mg/day			
			Intake	Urine	Stool	Balance	%	Intake	Urine	Stool	Balance
1	200	18	16.8	1.6	13.6	+1.6	19	930	637	203	+ 90
	200	30	17.2	1.5	13.8	+1.9	20	2004	1578	296	+130
2	800	18	13.6	0.8	13.5	-0.7	1	830	529	380	- 79
	800	36	13.1	0.7	12.0	+0.4	8	1981	1337	542	+102
3	2000	48	14.6	0.9	12.2	+1.5	16	937	443	472	+ 22
	2000	48	15.2	0.8	13.0	+1.4	15	2033	887	941	+205

intake of 2000 mg per day. The studies have shown that the addition of phosphorus to the diet did not change the net absorption of zinc during the different phosphorus intakes and this lack of change was independent of the calcium intake. Even during the concomitant use of a high calcium and a high phosphorus intake there was no change of the intestinal absorption of zinc and the net absorption of zinc was similar during the calcium intakes of 200 mg and 2000 mg per day.

The phosphorus balance data (Table II) show that the urinary phosphorus excretion increased approximately two- to three-fold in all three high phosphorus studies, while the fecal phosphorus excretion increased to a considerably lesser extent. The phosphorus balance of all three patients became more positive during the high phosphorus intake. These studies have shown that increasing the phosphorus intake up to 2000 mg/day did not affect the excretions of zinc nor the zinc balance, despite the fact that there were marked changes in phosphorus metabolism, primarily a marked increase of the urinary phosphorus excretion, while the fecal phosphorus excretion increased to a lesser extent.
This lack of change of the zinc excretions during the addition of phosphorus was observed irrespective of the calcium intake. Although it has been claimed frequently that the American diet has a high phosphorus content, it is not expected that the usual diet would reach or exceed 2000 mg phosphorus per day. This amount of phosphorus is higher than that officially recommended (17). It is noteworthy that in none of the studies in man reported here did the zinc balance become more negative during the high phosphorus intake in contrast to results obtained in animals (11). A review of the zinc requirement in man elaborates on various factors which influence the intestinal absorption of zinc, including the effect of phosphorus and of calcium (18).

Studies of the effect of both calcium and phosphorus on the zinc balance which have been carried out in this Research Unit (19, 20) have shown that the zinc balance varies during a normal dietary calcium and phosphorus intake and ranges from slightly negative to positive values and that the net or apparent absorption of zinc, calculated from zinc balance data, remained unchanged during the addition of the amounts of calcium or phosphorus used in these studies. The zinc balances in the present study have to be considered as maximal balances as the loss of zinc in sweat has not been considered as part of the excretory losses. It has been estimated that the loss of zinc in sweat is quantitatively as great as the urinary zinc excretion (21).

The results obtained in zinc balance studies during different intake levels of calcium and phosphorus were in agreement in ^{65}Zn absorption studies. Figure 1 shows data of ^{65}Zn plasma levels and of the fecal ^{65}Zn excretions following a single oral dose of ^{65}Zn chloride during the intake of different levels of calcium and phosphorus. The ^{65}Zn plasma levels were quite similar during the different study conditions as were the fecal ^{65}Zn

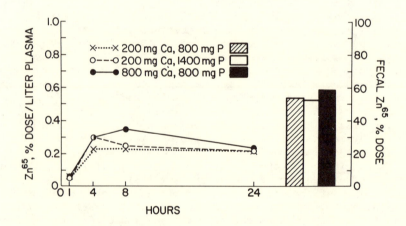

Figure 1. A single tracer dose of ⁶⁵ZnCl₂ was given orally in each study. Fecal ⁶⁵Zn excretions are cumulative for 12 days.

excretions, thus confirming the zinc balance data, i.e., the lack of effect of the amounts of calcium and phosphorus, used singly or combined, on the availability of zinc for absorption.

The results obtained with calcium and phosphorus on the zinc balance and on the availability of zinc for absorption in the present study are in contrast to results obtained in animals. These observations point out the difficulty in extrapolating results obtained in animal studies to humans.

Acknowledgment

This study was supported by a grant from the U. S. Department of Agriculture, CSRS, S&E.

Literature Cited

1. Berry, R. K.; Bell, M. C.; Grainger, R. B.; Buescher, R. G. J. Animal Sci. 1961, 20, 433–9.
2. Forbes, R. M. Fed. Proc. 1960, 19, 643–7.
3. Forbes, R. M.; Yohe, M. J. Nutr. 1960, 70, 53–7.
4. Hoekstra, W. G.; Lewis, P. K.; Phillips, P. H.; Grummer, R. H. J. Animal Sci. 1956, 15, 752–64.
5. Lewis, P. K.; Hoekstra, W. G.; Grummer, R. H. J. Animal Sci. 1957, 16, 578–88.
6. Newland, H. W.; Ullrey, D. E.; Hoefer, J. A.; Luecke, R. W. J. Animal Sci. 1958, 17, 886–92.
7. Bell, M. C.; Lloyd, M. K. Fed. Proc. 1963, 22, 492.
8. Oberleas, D.; Muhrer, M. E.; O'Dell, B. L. J. Animal Sci. 1962, 21, 57–61.
9. O'Dell, B. L; Savage, J. E. Proc. Soc. Exp. Biol. Med. 1960 103, 304–6.
10. Heth, D. A.; Hoekstra, W. G. J. Nutr. 1965, 85, 367–74.
11. Heth, D. A.; Becker, W. M.; Hoekstra, W. G. J. Nutr. 1966, 88, 331–7.
12. Spencer, H.; Vankinscott, V.; Lewin, I.; Samachson, J. J. Nutr. 1965, 86, 169–77.
13. Pecoud, A.; Donzel, P.; Schelling, J. L. Clin. Pharm. Therap. 1975, 17, 469–74.
14. Willis, J. B. Spectrochim. Acta 1960, 16, 259–72.
15. Willis, J. B. Analyt. Chem. 1971, 33, 556–9.
16. Fiske, C. H.; SubbaRow, Y. T. J. Biol. Chem. 1925, 66, 375–400.
17. "Recommended Dietary Allowances"; 9th ed.; National Academy of Sciences; Washington, DC, 1980.
18. Halsted, J. A.; Smith, J. C. Jr.; Irwin, M. I. J. Nutr. 1974, 104, 345–78.
19. Spencer, H.; Osis, D.; Kramer, L.; Norris, C. "Trace Elements in Human Health and Disease;" Prasad, A. S., Ed.; Academic Press, Inc.; 1976; Vol. 1, New York; pp. 345–61.

20. Spencer, H.; Gatza, C.; Kramer, L.; Osis, D. "Zinc in the
     Environment." Part II.; Nriagu, J. O., Ed.; John Wiley &
     Sons, Inc.; New York, 1980, pp. 105-19.
21. Prasad, A. S.; Schulert, A. R.; Sandstead, H. H.; Miale, A.;
     Farid, Z. J. Lab. Clin. Med. 1963, 62, 84-9.

RECEIVED August 23, 1982

Zinc Transport by Isolated, Vascularly Perfused Rat Intestine and Intestinal Brush Border Vesicles

MICHAEL P. MENARD, PAUL OESTREICHER[1] and
ROBERT J. COUSINS[1]

Rutgers University, Department of Nutrition, New Brunswick, NJ 08903

Studies of zinc transport with isolated, vascularly perfused rat intestine and brush border vescicles fail to support a role for a unique binding ligand in zinc absorption. Evidence suggests that a physiological pH, none of the ligands tested significantly enhanced zinc uptake or transfer by the perfused intestine. Uptake studies with brush border membrane vescicles also support these findings. In vescicles from both zinc normal and zinc depleted rats, zinc transfer does not require energy. With vescicles from normal rats, the Jmax is 5.4 nmoles per min, with a Km of 0.41 mM. The Km is not changed, but the Jmax increases to 12.0 nmoles per min with vescicles from zinc depleted rats. Electrophoretic data suggests a change in brush border membrane protein composition occurs during zinc deficiency. These proteins could influence zinc uptake during a reduction in dietary zinc supply.

The bioavailability of a mineral nutrient is influenced by a plethora of factors. Composition of the diet can influence the interaction between specific dietary components and metals. The physical characteristics of these interactions will determine how much mineral is available for absorption. These interactions are undoubtedly influenced by the acid conditions of the gastric secretions and the hydrolytic activity of the intestine. In order to understand how these factors might influence the absorption of a nutrient metal such as zinc, it is first necessary to understand the mechanism of zinc absorption and the systemic factors which regulate it.

[1]Current address: University of Florida, Department of Food Science and Human Nutrition, Gainesville, FL 32611

0097-6156/83/0210-0233$06.00/0

Twenty years ago Cotzias et al. (1) provided the first evidence for homeostatic control of zinc absorption. Subsequent reports from numerous laboratories have reinforced that hypothesis. Studies with both ruminants (2) and rats (3,4) have shown zinc absorption is depressed by high levels of dietary zinc. Conversely, zinc deficiency results in an increase in zinc absorption (5), with maximal absorption occurring within several days of depletion (6). The mechanisms which account for this homeostatic control have not been fully delineated.

The mechanism of zinc absorption involves several distinct phenomena. These include 1) uptake and transfer of zinc at the luminal membrane, 2) intracellular interaction between the intracellular zinc pool and newly absorbed zinc, 3) transfer of zinc at the basolateral membrane to the portal circulation and 4) transfer of zinc in a serosal to mucosal direction (7). The contributions of any number of these components and their interactions also influence the extent of zinc absorption and thus account for the wide range of bioavailability of zinc in various foods (8). While numerous studies on the influence of metabolic conditions on zinc absorption have been reported, there is at present no concensus on the specifics of this process. Earlier studies by Pearson et al. (9). Sahagian et al. (10) and Oberleas et al. (11) provided no consistent evidence for an energy requirement for the absorption of zinc. These studies conflict, however, with more recent efforts which support the concept of a requirement for active transport in some phase of absorption. Kowarski et al. found that 2,4 dinitrophenol, an inhibitor of ATP production, caused a decrease in zinc transport by rat jejunal segments suggesting an energy requiring process was involved (12). Similarly, Schwarz and Kirchgessner provided evidence that, in both zinc-depleted and pair-fed rats, zinc uptake in intestinal segments was inhibited by this compound (13). Since an energy requirement has been defined in liver parenchymal cells (14), it seems plausible that an energy requirement may exist for zinc absorption by intestinal cells. However, due to the functional polarity of the mucosal cell, the energy requirement may be at either the brush border membrane or the basolateral membrane. Definitive proof requires isolation of the individual membranes for separate study.

Uptake of zinc at the brush border membrane is undoubtedly influenced by association of the metal with various compounds in the intestinal lumen. A point of contention, however, has been the hypothesis that a unique zinc-binding ligand is a requisite for zinc absorption. Attempts to identify such a compound and establish a function have yet to be convincingly adduced. Nevertheless, a variety of compounds have been suggested as having a unique relationship to zinc absorption. Included are N,N,N'-trimethyl-1,2-ethanediamine (15), a peptide (16), prostaglandin E_2 (17), citric acid (18), picolinic acid (19), or amino acid residues (20). This subject has been reviewed recently by Cousins (21). Recent studies by Cousins et al. (22) have failed to provide evidence for a specific moiety involved in the absorption process. They have shown that the low molecular weight zinc binding ligands observed by various investigators can be explained as artifacts of the experimental protocol used. Intestinal preparations appear to be susceptable to proteolytic activity which can alter zinc binding profiles in mucosal cytosol. This results from atypical binding between intracellular components and degradation products. Similarly, Cherian noted rapid degradation in post-mitochondrial supernatants of mucosal cytosol which had been heated briefly (23). Further studies by Cousins et al. have shown that elevation of the intestinal zinc content to very high levels in vitro also yields anomolous binding characteristics. Apparently, exogenous zinc saturates higher molecular weight binding sites and atypical binding to lower molecular weight species results (22). Some of these findings have been confirmed by Lönnerdal et al. (24).

The second step in zinc absorption involves the intracellular interaction of zinc with various compounds which may enhance or impede absorptive processes. In 1969, Starcher noted that radioactive copper, given orally, associated with a low molecular weight protein (25). Subsequently, this mucosal protein was isolated and characterized by Richards and Cousins, who classified it as a metallothionein (26), and who further showed that it was induced in response to zinc administration (5). The appearance of this metallothionein, with properties similar to those described for both rat (27) and human (28) liver metallothionein, appears to be related to changes in both dietary zinc status and plasma zinc levels (5). The synthesis of mucosal metallothionein has been shown to be under transcriptional control (29,30). Menard et al. reported that dietary zinc administration resulted in enhancement of metallothionein mRNA transcription and its subsequent translation, to yield nascent metallothionein polypeptides (31). The intestinal metallothionein appearance was correlated to both an increase in mucosal zinc content primarily associated with the protein and with a decrease in serum zinc levels. In addition, Smith et al., using the isolated, vascularly perfused intestinal system, reported an inverse relationship between the synthesis of metallothionein and zinc transfer to the portal system, confirming earlier studies (32).

Factors involved in zinc transfer across the basolateral membrane to the circulation have not been well characterized. However, Smith and Cousins have suggested that the transfer of zinc across the serosal membrane may be the rate limiting step in zinc absorption and uptake to the circulation (33). A similar suggestion was made by Kowarski et al. (12). Zinc transport at the basolateral membrane appears to be a linear function of the lumen zinc content and directly related to the mucosal zinc lumen zinc content and directly related to the mucosal zinc concentration. Zinc transfer to the circulation requires albumin as the carrier protein (23). Transferrin does not appear to be significantly involved with portal transfer of zinc, although this has been proposed previously by Evans and Winter (34).

The nature and significance of endogenous zinc secretion into the intestine is not well understood. Secretion could occur as transcellular flux in the serosal to mucosal direction (12,35) and/or as pancreatic and other glandular secretions (36). Evidence in support of roles of the pancreas and liver in contributing to this pool of zinc is conflicting (37,38). In any event, endogenous secretion has been shown to be directly related to zinc intake at or above the dietary requirement (39). However, the physiological significance of this phenomena remains to be defined.

In recent experiments, we have employed two approaches to examine specific aspects of zinc absorption. The isolated, vascularly perfused rat intestine has been employed to examine absorption at the organ level. Isolated brush border membranes from rat intestine have been used to study the transfer of zinc across the intestinal surface.

The isolated, vascularly perfused rat intestine system has been used to investigate the influence of various zinc-binding ligands on zinc absorption. With this approach the functional integrity of the intestine with respect to several minerals, including calcium, iron and zinc uptake, and subsequent transfer of these minerals to their respective serum transport proteins is maintained (32,33). The intestinal perfusion system allows the simultaneous measurement of both mucosal zinc uptake (retention) and transfer to the portal circulation (absorption), and thus provides detailed information on the nature of the mechanisms of both uptake from the lumen and transfer to albumin in the portal circulation.

Male rats used for the perfusion experiments were maintained under established conditions and were fasted 16-24 hours prior to surgery as previously described (35). In the first series of studies the luminal perfusate was at pH 4.2. This perfusate consisted of M199 tissue culture medium which contained a variety of amino acids, vitamins and minerals plus glucose. The perfusate was supplemented (at 110 umolar) with L-histidine HCl, L-cysteine, L-methionine, L-tryptophan, 2-picolinic acid, citric acid, or reduced glutathione. The mixture was infused into the lumen at 0.39 ml per min for 20 min and 0.10 ml per min for the final 40 min of the experiments. The small intestine was then removed and mucosal cells harvested with a glass slide. The mucosa was homogenized and centrifuged. Aliquots of the post-mitochondrial supernatant and the portal effluent from the vascular perfusion were analyzed for ^{65}Zn content by gamma ray spectrometry. A second series of experiments was conducted in the same manner as above, but the pH of the luminal perfusate was adjusted to pH 6.6. The influence of 110 uM EDTA was also investigated at this pH. In a third series of experiments, the perfusate was supplemented with histidine, tryptophan, 2-picolinate, or citrate at 550 umolar.

When the pH of the luminal perfusate was at 4.2, only methionine had a significant (P < .05) effect on the amount of zinc appearing in the portal circulation. However, histidine and glutathione both significantly (P < .05) enhanced the amount of zinc sequestered in the intestinal mucosal cells.

As shown in Figure 1, when the luminal perfusate was adjusted to pH 6.6 to better simulate the intestinal environment, each of the compounds added to the perfusate appeared to decrease the amount of zinc transferred to the portal circulation. Histidine, cysteine, tryptophan and glutathione decreased transfer significantly. In marked contrast, however, EDTA was found to significantly enhance transfer to the portal supply. The retention of zinc by mucosal cells was significantly decreased by adding histidine, tryptophan, 2-picolinic acid and glutathione, as seen in Figure 2. EDTA did not alter zinc retention in the mucosal cells. Increasing the amount of histidine, 2-picolinate and citric acid in the lumen perfusate to a concentration of 550 uM had no effect on either mucosal retention or transmucosal flux of zinc. The same concentration of tryptophan, however, significantly decreased the amount of zinc taken up and transferred.

The results presented by these experiments fail to support the concept that a single low-molecular weight, naturally occurring species (ligand) uniquely enhances zinc absorption. The different compounds had various effects on both mucosal uptake and retention and portal transfer of zinc that were dependent upon both luminal pH and concentration of the ligand. In recent studies, Hurley et al. reported no effect of various ligands on zinc concentrations in rat dams and pups (40). Moreover, these perfusion data reaffirm the finding that EDTA enhances zinc absorption (41,42).

The uptake of zinc at the luminal surface and eventual transfer to the circulation is likely due to many factors acting together to influence zinc absorption. The affinity with which zinc is bound by a particular dietary component may influence its availability for uptake at the brush border membrane system. Alterations in zinc uptake noted under conditions of different lumenal pH may be due to the changing degree of ionization of various ligands, rendering the zinc more or less available for transport across the membrane.

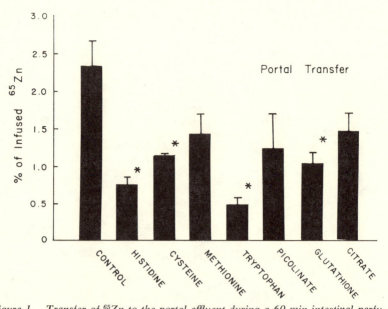

Figure 1. Transfer of ^{65}Zn to the portal effluent during a 60-min intestinal perfusion at pH 6.6, with zinc binding ligands added to the luminal perfusate. *Significantly different from control at P < 0.05.

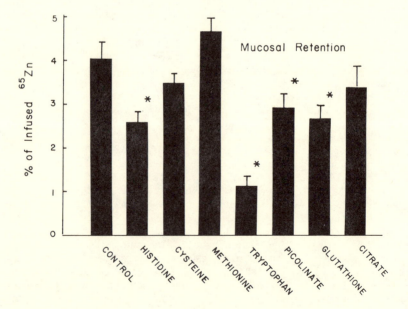

Figure 2. ^{65}Zn retention by mucosal cytosol following intestinal perfusion as in Figure 1. *Significantly different from control at P < 0.05.

In addition to factors influencing luminal uptake of zinc, transfer across the basolateral membrane has been shown to be dependent on the concentration of albumin in the portal circulation (33). These investigations suggest that metabolic factors which affect the albumin concentration in the plasma may also affect the rate of portal zinc transfer. It should be noted that EDTA did not enhance zinc accumulation within the mucosal cells yet it increased transfer to the vascular perfusate. These results suggest that basolateral membrane transport of zinc is enhanced by EDTA. We have proposed (35), as has Davies (38), that basolateral transport to the circulation is the rate limiting phase of zinc absorption. Since EDTA and zinc might be transported as a complex (42), the latter may transverse this barrier more easily and thus increase zinc absorption.

In addition to the vascular perfusion system studies, we have employed brush border membrane vesicles, isolated from rat mucosa, to determine more closely the parameters of mucosal zinc transport (43). These vesicle preparations represent the best means currently available to delineate the characteristics of zinc transfer into mucosal cells. The technique permits isolation of the microvillous membrane free of other cellular contaminants, as determined by established procedures (44).

For these experiments, brush border vesicles were obtained by a modification of the precipitation technique of Kessler et al. (45). Briefly, the small intestine was perfused via the mesenteric artery with ice cold 0.9% saline, the lumen rinsed with ice cold saline, and the mucosa scraped off with a glass slide. The mucosal tissue was suspended in a 50mM D-mannitol/10 mM Hepes/Tris buffer (pH 6.7), and homogenized with a blender for 5 min. Calcium chloride was added to a final concentration of 10 mM, and the resulting mixture was centrifuged at 3000 xg for 15 min at 4°C. The supernatant was recentrifuged at 42,000 xg for 20 min, and the pellet dissolved in the same buffer containing 10 mM EGTA, adjusted to pH 6.7. This mixture was centrifuged again at 42,000xg for 20 min and the pellet resuspended in the original buffer and recentrifuged to rinse the vesicles. The pellet was then suspended in the appropriate incubation buffer, homogenized with a Potter-Elvehjem homogenizer, and recentrifuged. The final pellet was then suspended with a 25 gauge needle in the appropriate buffer for the uptake studies. The above preparative steps were carried out at 4°C. Energy requirements for zinc transport were investigated by measuring zinc uptake under the influence of sodium and ATP equilibration. None was found, suggesting that an energy requirement, if one exists as has been proposed (12,13), must be at the basolateral membrane.

Once the requirements for optimal zinc uptake by brush border vesicles were defined, the parameters of zinc uptake under optimal conditions were determined for vesicles from both zinc normal and 4 day deficient rats. Rats were maintained on control diets with final zinc concentrations of either 1 or 50 ppm for periods up to 4 days. Experiments were initiated by adding zinc in extravesicular (luminal) concentrations varying from 34 uM to 450 uM to approximately 100 ug of vesicular protein in the appropriate buffer. Exactly one minute later, the reaction was terminated and vesicles collected on a Millipore filter for measurement of [65]Zn transfer.

The effect of some postulated zinc binding ligands on the uptake of zinc in vesicles was determined by adding either glutathione, citrate or 2-picolinate at an extravesicular concentration of 384 uM to the appropriate incubation buffer containing 192 uM zinc. Uptake was determined at various times up to 60 minutes. In order to determine whether the effect of the ligands on zinc uptake was specific, or due to a general effect on the membrane, the influence of the ligands on D-glucose uptake was analyzed. The ligands, in the above concentrations, were added to aliquots of vesicles containing 200 uM D-glucose, and uptake at various times up to 60 minutes were determined, as described above. The results of these experiments also fail to support a requirement for active transport at the brush border membrane. Uptake at 1mM and 192 uM zinc concentrations showed no evidence of either a Na or ATP requirement. Studies involving the influence of a 4 day zinc depletion on the same parameters yielded similar results.

Results from these studies support the findings of Davies (38), who suggested two possible pathways for zinc absorption at different luminal zinc concentrations. Our results at 192 uM zinc show a rapid uptake of zinc to saturation levels, while at higher (1mM) concentrations zinc uptake continued to increase throughout the experiment. These results are illustrated in Figure 3. When 1 minute zinc uptake was measured at extravesicular concentrations of from 34 uM to 450 uM, uptake increased in a curvilinear fashion towards a maximal rate of zinc uptake. Uptake at concentrations from 1 mM to 3 mM increased linearly with zinc concentration. When the uptake kinetics were presented as an Eadie-Hofstee plot, extravesicular zinc concentrations of 450 uM and below yielded a straight line characteristic of carrier mediated uptake, as seen in Figure 4. Calculations show that with vesicles from control rats, zinc uptake occurs with a Km of 0.38 mM and a Jmax of 5.4 nmoles per min. Values obtained with vesicles from 4 day zinc deficient rats showed no significant variation in Km, (0.44 mM), but the Jmax increased significantly to 12.0 nmoles per min. Statistical analyses of the kinetic data were performed (46). These data suggest that while half-maximal uptake was not altered in dietary zinc depletion, the uptake capacity of the brush border membrane increased dramatically.

Finally, to investigate the effect of zinc depletion on

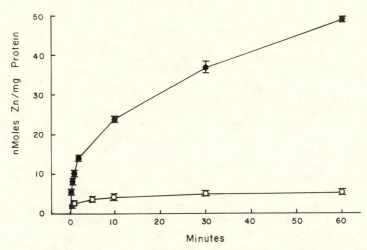

Figure 3. Zinc uptake by brush border membrane vesicles as a function of time under optimal (control) conditions at extravesicular zinc concentrations of 1 mM (●) and 0.192 mM (○).

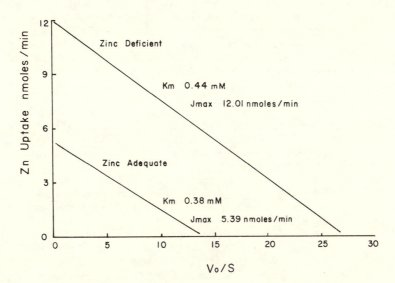

Figure 4. Eadie-Hofstee plot of 1 min zinc uptake by brush border membrane vesicles from control (+Zn) and 4-day zinc deficient (−Zn) rats.

the protein composition of the brush border membrane, vesicles were obtained from control, 4 day zinc deficient and 8 day zinc deficient rats. Equivalent concentrations of vesicular proteins were subject to SDS-polyacrylamide gel electrophoresis. As seen in Figure 5, zinc deficiency resulted in definite changes in brush border membrane protein composition. In zinc depletion studies, a general decrease in protein synthesis is believed to occur. However, when equivalent amounts of protein were subject to electrophoresis, there was a relative increase in the synthesis of at least two proteins. These had migrations equivalent to proteins of approximately 45,000 and 63,000 daltons. The role of these proteins in zinc uptake has not been established. However, since their appearance coincides with a significant increase in Jmax, questions as to their function in the membrane can be raised. It should be pointed out that the intensity of the 45,000 dalton protein band also increased in pair-fed animals, although not to the extent observed in the zinc depleted rats. The need for further studies on the relation of these proteins to zinc uptake is apparent.

Brush border membrane vesicles allow the determination of zinc uptake under precisely controlled conditions. When the effect of various zinc binding ligands on zinc uptake was measured, varying results were obtained. Although zinc uptake with picolinate and citrate at 60 min of incubation was significantly lower than control values, initial rates of zinc uptake did not vary significantly. These results support previous studies wherein these compounds did not alter zinc absorption (47). Moreover, when the effect of these ligands on D-glucose uptake was tested, no difference was noted. Therefore, it appears that the response noted above is specific for zinc uptake, possibly relating to the affinity of the ligands for zinc.

The studies reported here using the isolated, vascularly perfused rat intestine system and isolated brush border membrane vesicles fail to support a role for a specific zinc-binding ligand involved in zinc uptake in the rat. Rather, the extent of zinc uptake involves the interaction of several phenomena, including both extracellular and intracellular reactions. It appears that the major pathway of zinc uptake under normal dietary conditions involves the transfer of zinc from various dietary components to a carrier mediated transport system at the brush border membrane. The net absorption of zinc from the lumen could involve a competition between various dietary components, zinc binding ligands and the membrane carrier for zinc. Thus, in some cases, those compounds in the lumen with a higher affinity for zinc than the membrane component will be less likely to permit transfer of zinc to the carrier, while compounds with a lower affinity for zinc will increase the amount of zinc made

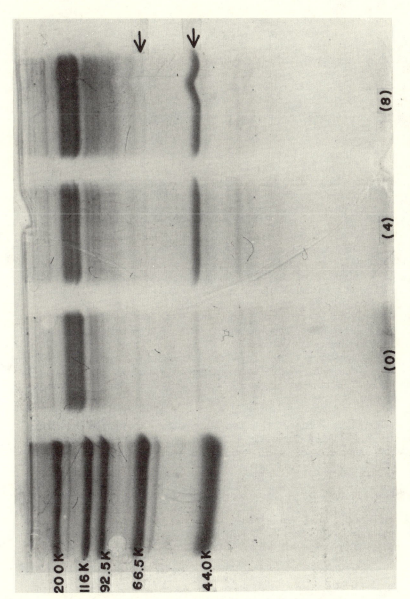

Figure 5. SDS-acrylamide gel electrophoresis of brush border membrane proteins from rats of adequate zinc status (0), following a 4-day zinc depletion (4), and following an 8-day zinc depletion (8). Proteins migrating with approximate molecular weights of 63,000 and 45,000 daltons are more apparent after 4 and 8 days of dietary zinc restriction.

available for transport. Some zinc-chelate complexes could be absorbed intact although there is little data on this aspect of zinc transport. Digestive action could alter these interactions and influence the availability of zinc for the membrane carrier.

Evidence for a similar membrane carrier for zinc in humans has been suggested from studies of zinc uptake in acrodermatitis enteropathica. In 1979, using mucosal samples obtained from acrodermatitis patients, Atherton et al. reported that, at lower lumenal zinc concentrations, brush border zinc uptake was inhibited in these patients, an effect overcome at higher zinc concentrations (48). As a result of these studies, they suggested the possibility of a membrane associated carrier for zinc in the brush border membrane. In the acrodermatitis enteropathica patients, this carrier could be altered or absent. Their hypothesis is supported by findings in rats by both Davies (38) and by this laboratory (43).

Based upon these and earlier studies, zinc absorption may be suggested to occur in the following steps. Under normal dietary conditions where an adequate supply of zinc is provided, initial uptake of zinc to the mucosal membrane appears to depend on its net availability for uptake to and transport by a membrane associated zinc carrier. This availability is, in turn, dependent on the zinc binding affinity of binding ligandsd and the membrane carrier. Zinc is transported into the mucosal cell, where it may either enter into intracellular reactions, including binding to proteins, newly synthesized thionein polypeptides (31) and/or binding to intracellular membrane components. The net flux of these reactions determines the availability of zinc for transfer at the basolateral membrane to albumin in the portal circulation. The albumin content of the plasma may also regulate this final phase of zinc absorption (33). Given the existence of ATPase in the basolateral membrane, the transport of zinc across this membrane may proceed by active transport. Several investigators have suggested zinc absorption involves active transport (12,13,39). However, further studies are needed to determine mechanisms of zinc transport across this membrane system.

Acknowledgments

The work from the author's laboratory (RJC) discussed in this review was supported by USPH, NIH Grant No. AM 18555 from the National Institute of Arthritis, Diabetes, Digestive and Kidney Diseases.

Literature Cited

1. Cotzias, G.C., Bong,C. and Selleck, B. Am. J. Phys., 1962, 202, 359.
2. Miller, W.J. Am. J. Clin. Nutr., 1969, 22, 1323.
3. Richards, M.P. and Cousins, R.J. J. Nutr., 1976, 106, 1591.
4. Becker, W.M. and Hoekstra, W.G. "Intestinal Absorption of Metal Ions, Trace Elements and Radionuclides. Skonyna, S.C. and Waldron-Edward,D., Eds.,Pergamon, Elmsford, NY, 1971; p.229.
5. Richards, M.P. and Cousins, R.J. Proc. Soc. Exp. Biol. Med., 1976, 153, 52.
6. Smith, K.T., Cousins, R.J., Silbon, B.L., Failla, M.L. J. Nutr., 1978, 108, 1849.
7. Cousins, R.J. Clinical, Biochemical and Nutritional Aspects of Trace Elements" Prasad, A.S., Ed., Alan R. Liss, Inc., New York, 1982.
8. O'Dell, B.L. Am. J. Clin. Nutr. 1969, 22, 1315.
9. Pearson, W.N., Schwink, T., Reich, M. "Zinc Metabolism" Prasad, A.S., Ed., Chas. C. Thomas, Springfield, IL, 1966, p.239.
10. Sahagian, B.M., Harding-Barlow, I., Perry, H.M., Jr. J. Nutr., 1966, 90, 259.
11. Oberleas, D., Muhren, M.E., O'Dell, B.L. J. Nutr., 1966, 90, 56.
12. Kowarski, S., Blair-Stanek, C.S. and Schacter, D. Am. J. Physiol., 1974, 226, 401.
13. Schwarz, F.J. and Kirchgessner, M. J. Anim. Phys. Anim. Nutr. Feedstuff. (Hamburg), 1977, 39, 68.
14. Failla, M.L. and Cousins, R.J. Biochim. Biophys. Acta., 1978, 538, 435.
15. Hahn, C., Severson, M.L. and Evans, G.W. Fed. Proc., 1976, 35, 863.
16. Hahn, C. and Evans, G.W. Proc. Soc. Exp. Biol. Med., 1973, 144, 793.
17. Song, M.K. and Adham, N.F. Am. J. Phys., 1978, 234, E99.
18. Hurley, L.S., Lönnerdal, B. and Stanislowski, A.G. Lancet, 1979, March 24, 677.
19. Evans, G.W. and Johnson, E.C. J. Nutr, 1980, 110, 1076.
20. Richards, M.P. and Cousins, R.J. Bioinorg. Chem., 1975, 4, 215.
21. Cousins, R.J. Am. J. Clin. Nutr., 1979, 32, 339.
22. Cousins, R.J., Smith, K.T., Failla, M.L. and Markowitz, L.A. Life Sci., 1978, 23, 1819.
23. Cherian, M.G. J. Nutr., 1977, 107, 965.
24. Lönnerdal, B., Keen, C.L., Sloan, M.V. and Hurley, L.S. J. Nutr., 1980, 110, 2414.
25. Starcher, B.C. J. Nutr., 1969, 97, 321.

26. Richards, M.P. and Cousins, R.J. Biophys. Biochem. Res.
    Comm., 1977, 75, 286.
27. Bremner, I. and Davies, N.T. Biochem. J., 1975, 149, 733.
28. Buhler, R.H. and Kagi, J.H. FEBS Letters, 1974, 39, 229.
29. Richards, M.P. and Cousins, R.J. Biochem. Biophys. Res.
    Comm., 1975, 64, 1215.
30. Richards, M.P. and Cousins, R.J. Proc. Soc. Exp. Biol.
    Med., 1977, 156, 505.
31. Menard, M.P., McCormick, C.C. and Cousins, R.J. J. Nutr.,
    1981, 111, 1353.
32. Smith, K.T. and Cousins, R.J. J. Nutr., 1980, 110, 316.
33. Smith, K. T., Failla, M.L. and Cousins, R. J. Biochem. J.,
    1979, 184, 627.
34. Evans, G.W. and Winter, T.W., Biochem. Biophys. Res.
    Comm., 1975, 66, 1218.
35. Smith, K.T., Cousins, R.J., Silbon, B.L. and Failla, M.L.
    J. Nutr., 1978, 108, 1849.
36. Pekas, J.C. Am. J. Phys., 1966, 211, 407.
37. Antonson, D. L., Barak, A.J. and Vanderhoof, J. A. J.
    Nutr., 1979, 109, 142.
38. Davies, N.T. Br. J. Nutr., 1980, 43, 189.
39. Weigand, E. and Kirchgessner, M. Nutr. Metab., 1976, 20,
    314.
40. Hurley, L.S., Keen, C.L., Young, H.M. and Lönnerdal, B.
    Fed. Proc., 1982, 41, (3) 781.
41. Kratzer, F.H., Allred, J. B., Davis, P.N., Marshall, B.J.
    and Vohrce, P. J. Nutr., 1959, 68, 313.
42. Suso, F. A., Edwards, H.M., Jr., Nature, 1972, 236, 230.
43. Menard, M.P., and Cousins, R.J. 1982, (Manuscript in
    Preparation)
44. Murer, H. and Kinne, R. J. Memb. Biol., 1980, 55, 81.
45. Kessler, M., Acuto, O., Storelli, C., Murer, H., Muller,
    M. and Semenza, G. Biochim. Biophys. Acta, 1978, 506, 136.
46. Cleland, W.W. Adv. Enzymol., 1967, 29, 1.
47. Oestreicher, P., Menard, M.P. and Cousins, R.J. Fed. Proc.,
    1981, 40, (3), 937.
48. Atherton, D.J., Muller, D.P.R., Aggett, P.J. and Harries,
    J.T. Clin. Sci., 1979, 56, 505.

RECEIVED October 25, 1982

Competitive Mineral—Mineral Interaction in the Intestine

Implications for Zinc Absorption in Humans

NOEL W. SOLOMONS[1]

Massachusetts Institute of Technology, Department of Nutrition and
Food Science, Cambridge, MA 02178

Competition with other minerals may reduce the
biological availability of dietary zinc. Studies
in experimental animals describe biological mineral-
mineral interactions between and among metals of
similar chemical configurations. Chemically-similar
nutrients could theoretically reduce zinc absorption by interfering with: 1) its uptake into the
cell; 2) its transfer across the cell; and/or 3) its
transport into the circulation. Competitive interactions between zinc and copper, iron, cadmium and
tin have been demonstrated in animals. In human subjects, high zinc levels can reduce copper absorption;
the converse has not been demonstrated. Fe/Zn ratios
of 2:1 or greater reduce the absorption of dietary
zinc. In the formation of vitamin-mineral supplements
and infant foods, adverse nutritional consequences
from an Fe:Zn imbalance could result. Sn/Zn ratios
of about 4:1 reduced the apparent absorption of zinc
in human subjects. Storage of food in tin-plated
cans might compromise zinc availability. Calcium and
and magnesium have no competitive effect on zinc
absorption in humans based on available evidence.

This symposium testifies to the concern among nutritional
scientists and food technologists that the biological availability
of dietary zinc may be a major determinant of zinc nutriture in
humans. Those assessing dietary adequacy with respect to a nutrient
such as zinc, must be concerned with two features: the density
(metallo-caloric ratio) and bioavailability.

The Food and Nutrition Board of the National Academy of
Sciences has established levels of recommended daily intake for
healthy individuals of various age-groups, sexes and reproductive
states: for infants 0-6 mo, 3 mg; for infants 7-12 mo, 5 mg; for
children 1-10 yrs, 10 mg; for adults, 15 mg; for pregnant women,
20 mg; for lactating women, 25 mg (1). More refined estimates

[1] Current address: Institute of Nutrition of Central America and Panama, Division of
Human Nutrition and Biology, Guatemala City, Guatemala, Central America

0097-6156/83/0210-0247$06.00/0

for young children, including premature infants, have been developed by Hambidge and Casey (2). The limited survey data on customary zinc intakes in the United States (3) suggest that most individuals consume from 45 to 70% of the RDA for zinc.

This low intake results from the relatively low total energy consumption of the sedentary U.S. population and the relatively low density of zinc in the mixed U.S. diet. To achieve the RDA for zinc with a 3000 kcal daily intake, dietary zinc density would have to be 5-7.5 mg/1000 kcal. The average mixed diet has a zinc density in the 3-4 mg/1000 kcal range (4).

The amount of zinc that must be absorbed daily, i.e., the parenteral requirement for maintenance of normal zinc balance, is considerably less than the RDA levels. Based on the estimate of an Expert Committee of the American Medical Association for the parenteral dosage of zinc in stable adults (5), absorbing anywhere from 17 to 27% of the RDA levels would provide the zinc needed for nutritional maintenance. Thus, the efficiency of absorption of dietary zinc can be an even greater determinant of adequacy than the dietary zinc density or daily total zinc intake, per se. With a focus on biological availability of zinc in human nutrition, the present paper reviews conceptual and experimental information from many laboratories, including our own, and draws conclusions about the possible impact of mineral-mineral interactions as factors affecting zinc bioavailability from human diets. The calcium:zinc interaction has been addressed by Dr. Spencer in the preceding chapter and will not be discussed here.

Biological Interactions of Chemically Similar Minerals

Interest in mineral-mineral interactions emerged in the 1960s. Most early experiments in laboratory animals involved either radioisotopic tracers or manipulation of the dietary ratios between and among minerals. In the latter feeding experiments, the exacerbation of manifestations of deficiency of one mineral (M_1) by excess amounts of a second (M_2), or the amelioration of the clinical signs of M_2 toxicity by increasing the dietary levels of

Table I. Interaction of zinc and copper on hemoglobin concentration, mortality, and body weight of chicks

Zinc, ppm	Copper, ppm					
	0	10	0	10	0	10
	Hb, g/100 ml		% Mortality		Wt, g	
0	6.5	8.5	8.7	0	262	272
50	5.9	8.0	18.2	4.5	205	320
100	6.0	8.8	52.0	7.7	182	294
200	4.8	8.0	70.8	20.0	139	291
300	4.1	6.6	88.0	4.2	127	310

after Hill and Matrone, 1970 (6)

M1 were the standard outcome variables to demonstrate a competition between two minerals. A typical example of an experiment in chicks concerning the interaction of zinc and copper is shown in Table I.

By 1970, enough data had been gathered for Hill and Matrone to formulate their classical, conceptual treatise, entitled: "Chemical parameters in the study of in vivo and in vitro interactions of transition elements" (6). The authors' thesis was: "Those elements whose physical and chemical properties are similar will act antagonistically to each other biologically" (6). Thus, as demonstrated in Table I, increasing levels of dietary zinc interfered with the role of copper in hematopoiesis, survival and growth in copper-deficient chicks. The authors related the mechanism of biological mineral-mineral antagonism to the similarities of the electron orbital configurations of the ions. The orbital arrangements in the outer shells of cupric (Cu^{++}) and cuprous (Cu^{+}) ions are shown in Fig. 1. Noting that Zn^{++}, Cd^{++} and Hg^{++} all had the same electronic structures in the valence shells as cuprous ion, and Ag^{++} the same as cupric ion, they predicted that copper would show biological interaction with each of the divalent ions listed. Review of the literature and experiments in their own laboratory confirmed the hypothesis in all instances, except that of a copper:mercury interaction. It has since been established that other physicochemical factors account for some of the biolgical recognition of chemically-similar ions. At some level of avian or mammalian systems, mineral-mineral interactions involving zinc have been identified with iron, copper, cadmium, magnesium, calcium, phosphorus, cobalt and tin. The present review deals exclusively with intestinal competition.

As noted, the conventional experimental approach involves the use of dietary extremes, with levels of one or another nutrient that will induce deficiency or toxicity. Obviously, mineral-mineral ratios occurring in human diets or in the formulation of nutritional supplements bear on the implications of mineral competitions for human nutrition. Thus, in extrapolating from animal data or in designing and interpreting human experiments, the appropriateness of the mineral-mineral ratios to human nutrition is paramount.

Mechanistic Possibilities for Intestinal Interations

In discussing the absorption of minerals, especially in the context of their biological availability, it is important to clarify three terms for the subdivisions of the absorptive process: uptake refers to the entrance of the nutrient into the mucosal cell from the intestinal lumen; transfer refers to the passage of the nutrient out of the mucosal cell; and transport refers to the removal of the nutrient from the vicinity of the intestine to other parts of the organism. All three processes are essential for a dietary component, such as zinc, to achieve a nutritional role for the the organism. From the array of mechanisms thought

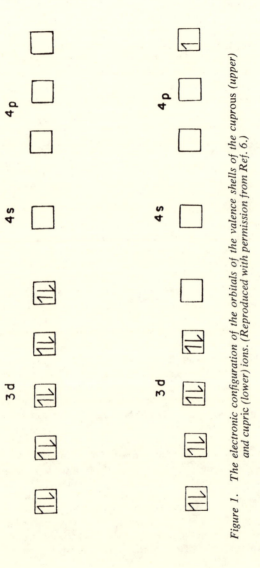

Figure 1. The electronic configuration of the orbitals of the valence shells of the cuprous (upper) and cupric (lower) ions. (Reproduced with permission from Ref. 6.)

to be involved in the absorption (uptake, transfer, transport) of dietary zinc, we can postulate various mechanisms of interaction of zinc with other minerals in the intestine. These are schematically illustrated in Fig 2.

Intraluminal Interactions

Direct mineral–mineral interaction (A). Zinc and another mineral could interact directly, forming a complex with one another, or together with a third moiety that alone would complex neither. Two minerals could also interact in oxidation:reduction reactions, in which the preferred oxidation state for absorption is altered for one (or both) minerals, e.g., $M_1^{++} + M_2^{++} \longrightarrow M_1^+ + M_2^{+++}$. The stability of the divalent state of zinc is such that it acts neither as an oxidizing nor as a reducing agent under the physiological conditions of the mammalian gastrointestinal tract.

Competition for Common Binding-ligand (B) or Complexing Agent (C). Zinc and another mineral could compete for a compound that acts as a binding-ligand favoring the zinc absorption. Excess of the competing mineral would displace zinc from the ligand and reduce its uptake. Alternatively, another mineral could compete with zinc for a complexing substance which forms an insoluble complex. Displacement of zinc from this form of binding would favor its biological availability.

Mucosal Interactions

Competition for a Common Receptor Site (D). Zinc and another mineral could compete for a common receptor mechanism, involved in the uptake of both metals across the mucosal membrane. High concentrations of the competitor would reduce the transport of zinc from the intestinal lumen into the cell.

Competition for a Common External Carrier-Protein (E). A variant of this aforementioned mucosal interaction would be a competitive interaction of zinc with another mineral for an externalized carrier-protein that might act as a shuttle to translocate metals across the membrane into the cell.

Intracellular Interactions

Competition for a Common Internal Carrier-Protein (F). Zinc and another mineral could also compete within the cell for a common carrier mechanism essential to the transport of zinc toward the basolateral portion of the cell.

Competition for a Common Intracellular Binding-Protein (G). Cousins (7,8) and others have suggested that intestinal metallothionein (MT), a sulfur-rich protein of 6,000–10,000 daltons, acts as a trap, capturing the zinc that enters the cell from the lumen (or from the bloodstream) and holding it until the zinc reenters the fecal stream with the normal exfoliation of intestinal mucosa. Since MT binds other minerals beside zinc (notably, cadmium, copper and mercury), simultaneous intake of the aforementioned minerals could result in less retention of zinc within the mucosal cell.

Induction of Intracellular Binding-Proteins (H). The steady-state concentration of an intracellular binding-protein such as

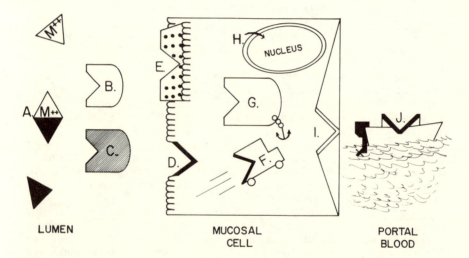

LUMEN   MUCOSAL   PORTAL
                              CELL   BLOOD

Figure 2.   Scheme of potential loci of mineral–mineral interactions between an index nutrient M^{2+} ($\triangle$) and another similar mineral ($\blacktriangle$). Sites include the lumen (A–C); the mucosal membrane (D, E); within the intestinal cell (F–H); and at the serosal surface, either on the membrane (I) or in the portal blood stream (J).

zinc between a metabolically available and metabolically unavailable pool can be a determinant of absorption. Various minerals (copper, cadmium, zinc) can induce the synthesis of MT (9,10). A habitual diet rich in these minerals might raise MT levels and establish conditions for a reduced zinc transfer.

Serosal Interaction.

Competition for a Common Serosal Receptor (I). The basolateral aspects of the intestinal cell represent a barrier to movement of a nutrient, at times as formidable as the brush-border membrane. The possibility of a competitive interaction for exit from the cell is also worth considering.

Competition for Transport on a Circulating Carrier-Protein (J). Once across the serosal border, zinc must still be bound by a transport protein to carry it through the portal circulation, away from the intestine. There is debate as to whether this role is played by transferrin (11) or by albumin (12,13). In either case, the presence of excess metal(s) that bind to the same site on the portal transport protein as zinc would impede the passage of zinc into the systemic circulation.

Copper:Zinc Interactions

Perhaps the most studied of biological mineral–mineral interactions involving zinc is the copper:zinc interaction. It was early noted that the anemia induced by toxic amounts of zinc (levels from 0.75 to 1.0%), could be ameliorated by adding copper (14-17). The depression of cytochrome C oxidase induced by excessive zinc could also be reduced by adding supplemental copper to the rations (15, 17,18). More recently Magee and Grainger (19) have shown that zinc contents of 50 and 100 ppm reduce liver copper concentrations in young rats. In the Japanese quail fed a diet marginal in copper, and containing 250 ppm zinc (Zn/Cu ratio ~ 250:1), decreased tissue copper and more severe signs of copper deficiency were observed (19). In guinea pigs fed a ration containing 2 ppm copper, dietary zinc levels of 11, 48 and 95 ppm produced hair copper concentrations of 8.9, 4.7 and 3.3 µg/g respectively (20).

Studies by Cox and collaborators showed that a 0.4% zinc enrichment of maternal rat rations during gestation (21,22,23) or lactation (24,25,26) would reduce copper content and cytochrome C oxidase activity in the tissues of fetuses and pups and the copper content of the milk (25). Adding 0.2% copper to the diet reversed this effect. Whether the zinc:copper interaction is mediated exclusively at the maternal gut, or at other foci of mineral transport in the dam or offspring cannot be determined.

Using radiocopper (64Cu) tracers, Van Campen (27), Van Campen and Scaife (28) and Evans et al. (29) reported inhibition of zinc transport with Zn/Cu ratios of 500:1 to 1000:1, but Ogiso et al. (30) found a significant inhibition of 64Cu absorption with a ratio of 30:1. Unique among studies in this area is the report by Van Campen (31) of the effect of excess copper on zinc (65Zn)

absorption. Five µg of zinc, labeled with ^{65}Zn in 0.5 ml of
distilled water, were introduced into intestinal segments of the
rat, with or without 200 µg of copper as cupric nitrate. The mean
transfer of radiozinc to the blood, heart, kidney and liver in 3 h
was 19%, but fell to 7.5% with the 50:1 Cu/Zn ratio. The disap-
pearance of zinc from the intestinal segments was 40.6% with zinc
alone, and 28.3% with intraduodenal copper.

Subsequent studies have given some insight into the cellular
mechanisms of the copper:zinc interaction. Evans et al. (29)
found the 1000:1 Zn/Cu ration inhibited ^{64}Cu uptake only in zinc-
deficient rats, and that its effect was not on uptake from the
lumen, but rather on transfer to the body. Parenteral injection
of copper prior to luminal dosing with ^{65}Zn also reduced net
copper absorption (32) (Fig 3) but again, this was due to reduced
transfer of copper across and out of the mucosal cell. Also, in
the study of Van Campen (31) of the effect of excess copper on
zinc absorption, indications of a predominant influence within
the mucosal cell were developed.

These observations led to the proposition that an intracellular
binding-protein, specifically MT, was involved in the absorption
of copper, and that this protein was the template for the
zinc:copper interaction (33,34,35). Fischer et al. (36) studied
the uptake and transfer of stable copper across everted intestinal segments in rats fed a range of zinc from 7.5 to 240 mg per
kg of diet. With increasing dietary zinc intake, a graded
increase in copper uptake from the lumen and in copper retained
in the mucosal cell, and a graded decrease in the transfer of
copper from mucosal cell to the interior of the sac was observed.
The intracellular copper was bound in the cytosol to a protein
with the molecular weight of MT. The authors suggest than zinc
exerts its antagonistic effect by inducing the synthesis of a
copper-binding protein (thionein) which sequesters copper within
the cell, preventing its transfer to the serosa.

Three human metabolic studies have examined the effect of
dietary zinc intake on apparent copper absorption. Greger et
al. (37) studied 11 adolescent girls, ranging in age from 12.5 to
14.2 years, during a 30-day balance period. The copper intake was
1.2 mg, all provided from natural foodstuffs. Total zinc intakes
were 11.5 or 14.7 mg/day, achieved by adding 4.4 or 7.6 mg of zinc
sulfate solution to the lemonade served at lunch. The remaining
foods provided 7.1 mg of zinc. Thus, Zn/Cu ratios of 9.6 and
12.2 were achieved in the respective experimental regimens. Each
level of zinc was fed for 15 days. Fecal loss of copper was 0.79
+ 0.10 mg/day (mean + SD) on the lower zinc intake and 0.90 + 0.16
mg/day on the higher intake. These copper outputs were significantly different (p <0.041).

Burke et al. (38) studied 11 geriatric subjects, 5 males and 6
females, ranging in age from 56 to 83 years, in a similar 30-day
metabolic study. Daily copper intake was 2.33 mg, with 1.05 mg
provided as copper sulfate added to the breakfast beverage and

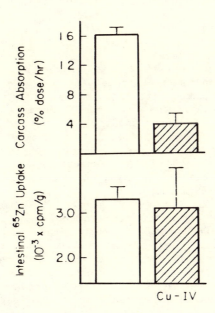

Figure 3.   Transfer to the carcass (upper) and the content in the mucosal lining (lower) of a gavage dose of ^{65}Zn administered to two groups of rats: untreated controls (open bars) and rats pretreated with an i.v. dose of copper (stippled bars). (Reproduced with permission from Ref. 32. Copyright 1975, The American Physiological Society.)

the remaining 1.28 mg from the foods. Daily zinc intakes were either 7.80 or 23.26 mg, with 6.03 mg provided by the diets and the remainder as zinc sulfate in the morning beverage. The daily Zn/Cu ratios were 3.3 and 10.0 respectively. The amount of zinc and copper in the morning beverages on the respective diets, however, was 1.77/1.03 mg and 17.23/1.03 mg. Five individuals received the low-zinc regimen and six individuals the high-zinc regimen. Mean fecal copper excretion values were 1.40 ± 6.8 mg (mean ± SD) and 2.03 ± 0.32 mg per day on the respective diets. The apparent copper retention on the low-zinc regimen, 0.94 ± 0.68 mg/day, was significantly greater than that on the high-zinc regimen, 0.30 ± 0.32 (p <0.05).

Taper et al. (39) from the same laboratory as the aforementioned geriatric study, found no significant effect of dietary zinc levels when they studied 18 young women of childbearing age. Dietary copper intake was 2.0 mg day with 0.86 mg from food and 1.14 mg as supplement. Foodstuffs provided 5.63 mg of zinc, and total daily intakes were adjusted to 8 mg (N = 7), 16 mg (N = 6) and 24 mg (N = 5). The experimental details of the supplementation with zinc and copper salts were not provided. The dietary Zn/Cu ratios were 4.0, 8.0, and 12.0 respectively; the mean fecal excretion (mean ± SD) for the respective diets was 2.01 ± 0.15, 2.10 ± 0.14 and 2.02 ± 0.23, respectively. Neither the fecal outputs nor the derivative apparent retention values differed significantly one from another.

Implications for Human Nutrition. The human studies cited have examined the effect of zinc on copper uptake, and evidence suggests that Zn/Cu ratios of greater than 10:1 can produce a nutritional impact. The source of the discrepancies among the three metabolic studies reviewed is not immediately evident. One possibility would be a different susceptibility to copper malabsorption at different ages. Another might have to do with the actual Zn/Cu ratios in those meals which provided the bulk of the daily intake of the two minerals or with their chemical forms. Studies in which all of the copper and zinc are derived from natural foodstuffs would complement the foregoing human studies in this area.

In addition, megadoses of zinc of up to 5 g per day (40) or chronic dosing with "therapeutic" amounts of zinc, 150 mg of zinc daily for 1 to 2 years (41), have been associated with overt copper deficiency anemias.

There are few circumstances in which a reversal of the ratio -- that is with a Cu/Zn ratio approaching or exceeding unity -- would occur. Certain vitamin-mineral preparations, e.g., Geriplex®, geriatric vitamin-mineral formula (Parke-Davis, Morris Plains, NJ), with 14 mg of copper and 2 mg of zinc, will provide such Cu/Zn ratios. Inhibition of zinc uptake by excess copper, as shown in rats by Van Campen (31), might affect human nutrition when Cu/Zn ratios are greater than one; direct confirmation is presently lacking, however.

Iron Zinc:Interaction

Biological interactions between iron and zinc were first suggested in 1960. Cox and Harris (42) found reduced hepatic iron in rats fed a high-zinc diet (0.4%), and Magee and Matrone (17) observed a differential distribution of absorbed radioiron (59Fe) in rats on a still higher zinc diet (0.75%); zinc supplementation did not influence iron absorption in this latter study. More recently, however, Magee and Fu (43) observed an apparent iron absorption of 42% in rats fed a control diet, and 0 absorption with a ration containing 0.75%.

Pollack et al. (44) demonstrated a significant 42% increase in the absorption of radiozinc (65Zn) tracer by anemic rats fed an iron-deficient diet. Iron deficiency induced by bleeding, however, does not enhance zinc uptake (44,45,46). Using isolated intestinal segments of mice fed either iron-adequate or iron-deficient diets, the absorption of zinc in the perfusate was, once again, found to be enhanced (47). Adding graded amounts of iron to the perfusates, to produce Fe/Zn ratios of 2.5, 5 and 10, caused a progressive decrease in the uptake and transfer of 65Zn by the intestine. Conversely, an 8:1 Zn/Fe ratio induced the uptake and transfer of radioiron (59Fe) in the intestine of iron-deficient mice.

Our laboratory became interested in the question of the physiological and nutritional importance of competition between iron and zinc while investigating the impact of various public health strategies for improving the iron nutriture of populations (48,49). We have since explored the iron:zinc interaction in human subjects, using the change in plasma zinc concentration after oral ingestion of pharmacological doses of zinc. In our initial studies, 25 mg of zinc was administered in 100 ml of CocaCola®. (CocaCola was used to mask the strong and unpleasant metallic taste of the salts.) Zinc was given alone, or with 25, 50 or 75 mg of ferrous iron, constituting Fe/Zn ratios of 0, 1:1, 2:1 and 3:1. As shown in Figure 4, a progressive reduction in the area under the discontinuous curve of plasma zinc concentration was produced (50). When a 3:1 Fe/Zn ratio was provided using heme iron from heme chlorided as the source of iron, no interaction was noted, nor was any inhibition of the uptake of 54 mg of zinc from fresh oysters seen when 100 mg of iron as ferrous sulfate -- 2:1 Fe/Zn ratio -- was administered (50). According to the Hill and Matrone dictum (6), a more chemically-disimilar ion, i.e., trivalent, ferric iron, would have a lesser effect on zinc absorption. The rise in plasma zinc from a 2:1 mixture of 50 mg of ferric iron and 25 mg of zinc was intermediate between zinc alone and a 2:1 Fe/Zn ratio with ferrous sulfate.

By using current theories about the mechanism of iron absorption, and manipulating the conditions related to iron, we have probed the mechanism involved in the interaction of iron and zinc. We reasoned that if the iron:zinc competition were in the lumen, then increasing the iron uptake would diminish the plasma

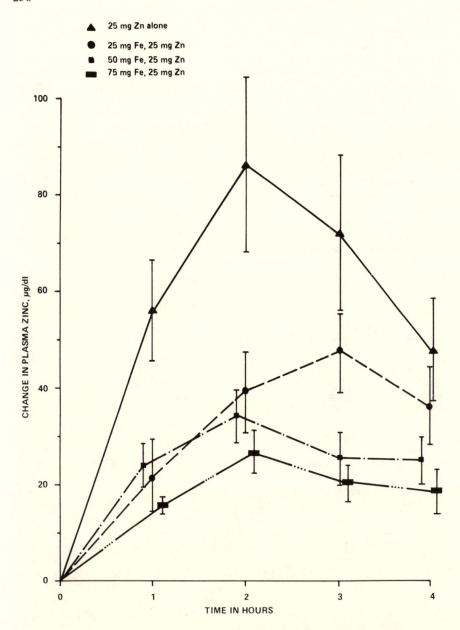

Figure 4.   The mean increment in plasma zinc concentration (± SEM) at hourly intervals after 7–8 subjects per group received zinc sulfate alone or with ferrous sulfate in an aqueous solution of CocaCola with the amounts of minerals shown. The distinct treatments constituted Fe/Zn ratios of 0, 1:1, 2:1, and 3:1. (Reproduced with permission from Ref. 50. Copyright 1981, American Society for Clinical Nutrition.)

appearance of zinc (Fig 5). When 1 g of ascorbic acid was added to the 2:1 solution of ferric iron and zinc, zinc absorption was inhibited; the conformation of the plasma zinc curve was now identical to that with a 2:1 Fe/Zn ratio with ferrous iron. The location of the interaction, however, could still be intracellular.

If the interaction of iron and zinc were functionally proximal to the site of the regulation of iron by the state of iron reserves of the host, then differences in iron nutriture would not affect the uptake of zinc in the presence of iron (Fig 6). We found an inverse correlation with a correlation coefficient of -0.546 (p <0.05) between the increment in circulating zinc after ingesting a 2:1 ferrous iron:zinc solution and plasma iron concentration. Thus, it would appear that the interaction of iron and zinc occurs <u>after</u> the intestinal regulation of iron.

The acute effect of daily administration of therapeutic amounts of iron is the reduction in the post–iron elevation in iron concentration. This has been seen in preschool children receiving 6 mg of Fe/kg body weight on consecutive days, but not alternate days (<u>51,52</u>). If the interaction of iron and zinc were <u>at</u> the site of this blockade, we would expect everyday treatment with therapeutic doses of iron to reduce the plasma uptake of zinc (Fig 7). The mean change in plasma zinc concentration afer a 25 mg oral dose of zinc in 5 volunteers who ingested 130 mg of iron daily for 4 consecutive days before the absorption test was identical to that of subjects with no pretreatment. Having established that the iron blockage, <u>per se</u>, does not affect zinc absorption, we reasoned that if the zinc:iron interaction were "distal" to the site of the iron blockade, then consecutive-day --but not alternate day -- dosing with therapeutic iron would decrease the inhibition of zinc absorption from a test solution containing a 2:1 Fe/Zn ratio (Fig. 8). Comparison of the rise in plasma zinc produced after an oral dose of a solution of 50 mg of iron and 25 mg of zinc showed no differences whether subjects received no pretreatment, 4 days of therapeutic iron (130 mg/day) or therapeutic iron 2 and 4 days before the absorption test.

Data from our studies involving manipulations of iron are consistent with an interaction of zinc and iron in the human intestine both in the lumen and at some intercellular location distal to the site of regulation of iron absorption by the iron status of the individual.

<u>Implications for Human Nutrition.</u>

To what extent does the interaction of iron and zinc influence, adversely, zinc nutriture in a dietary context. Natural foods tend to contain an iron:zinc ratio between 0.5 and 2.0 (Table II). Occasionally, the ratio of these two elements exceeds this range, as in the case of an iron-fortified cereal such as Cheerios. It is likely, however, that when a range of foods is blended into a meal, the Fe/Zn ratio will be close to unity.

Given the RDAs established for zinc and iron for various age-groups (<u>1</u>), if a person consumes his or her respective intakes of

Figure 5.   Scheme of the consequences of an iron:zinc competition at the mucosal membrane. Increasing the intraluminal concentration of iron (below) would lead to competitive inhibition of the entry of zinc into the cell.

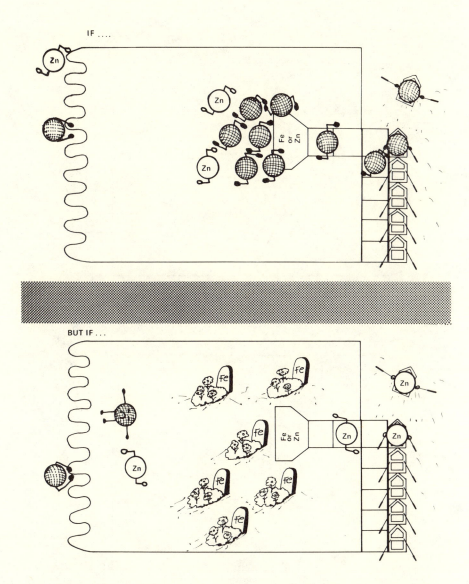

Figure 6.   Scheme of the consequences on iron:zinc interaction of a "mucosal block" regulation of iron transfer. If the site of iron:zinc competition is distal to the site of iron regulation (above) then iron sufficiency should reduce iron:zinc competition, increasing net zinc uptake.

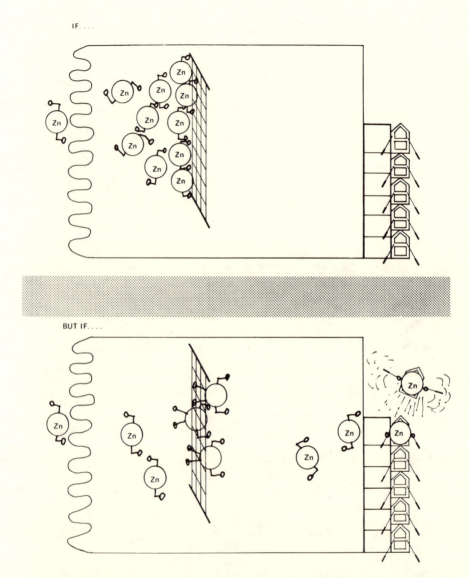

Figure 7.   Scheme of the consequences on iron:zinc interaction whereby zinc would either be blocked (above) or uneffected (below) by acute oral administration of iron.

IF....

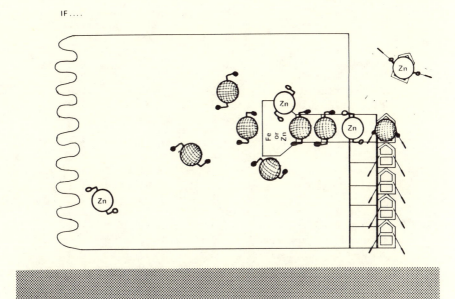

THEN...

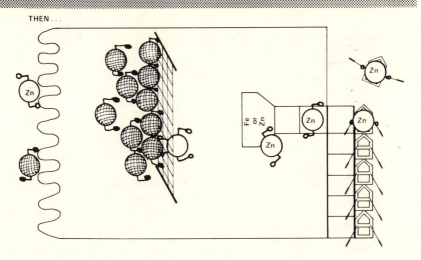

Figure 8. Scheme of the consequences on iron:zinc interaction of an acute blockade with therapeutic oral doses of iron. If the site of iron:zinc competition is distal *to the site of acute blockade of iron uptake (above), then oral administration should reduce iron:zinc competition, increasing net zinc uptake.*

Table II.   Iron and Zinc Content of Selected Foods
            and Beverages

	Fe mg/100 g	Zn mg/100 g	Fe/Zn ratio
Cheerios	16.0	2.54	6.3
Boiled lentils	2.1	1.00	2.1
Puffed rice cereal	2.8	1.42	2.0
Pineapple juice	0.3	0.16	1.9
Oatmeal cookies	2.2	1.33	1.7
Chocolate cake/ Chocolate icing	1.0	0.68	1.5
Avocado	0.6	0.43	1.4
Frozen pizza	1.3	1.22	1.1
Creamed spinach	0.7	0.61	1.1
Prepared horseradish	1.1	1.05	1.0
Cheese ravioli, frozen	1.4	1.52	0.9
Cooked brown rice	0.5	0.61	0.8
Creamed cottage cheese	0.3	0.46	0.6
Boiled cabbage	0.3	0.87	0.3
Parmesan cheese	0.7	5.50	0.1

Reproduced with permission from Ref. 53. Copyright 1980, Lippincott.

the elements, the Fe/Zn ratio will be less than 2.0 (Table III).
The notable exception to this rule is the infant. For babies from
birth to 6 months, the respective RDAs would constitute an Fe/Zn
ratio of 3.33. For the second semester of infancy, the corresponding Fe/Zn ratio is 3.0. These are levels at which significant
and substantial inhibition of zinc uptake was seen with ferrous
iron and inorganic zinc (50). If commercial infant formulas are
made to the specifications of the RDA, there could be harmful
effects on zinc nutriture. Such a case of iron:zinc interaction
might have, indeed, been demonstrated by Walravens and Hambidge
(54). These authors conducted a longitudinal study in which a
group of infants from white, middle-class homes in Denver,
Colorado were randomized at birth to receive Similac with iron
(Ross Laboratories, Columbus, OH), either in the original form
(1.8 mg zinc per liter) or supplemented with zinc sulfate (5.8 mg
zinc per liter). Since the infant formula was iron-fortified, it
contained 12 mg of iron per liter. Thus, the Fe/Zn ratios were
6.0 and 2.0, respectively. The male infants in the zinc-
supplemented cohort had lesser gains in height (2 cm) and weight
(535 g), and a 12% lower plasma zinc level after the 6-month
trial. Clearly, there was a three-fold difference in zinc content
of the formula, but the large Fe/Zn ratio might have been another
factor in the differential growth performance and zinc nutriture
of the children (50).

Table III. Recommended Dietary Allowances for Zinc
and Iron.

	Age (yrs)	Iron Allowance	Zinc Allowance	Fe/Zn Ratio
Infants	0.0–0.5	10	3	.33
	0.5–1.0	15	5	.0
Children	1–3	15	10	1.5
	4–10	10	0	1.0
Males	11–18	18	5	1.2
	19+	10	5	0.67
Females	11–50	18	5	1.2
	51+	10	5	0.67
Pregnant		18+	20	0.9+
Lactating		18	25	0.72

Reproduced with permission from Ref. 1. Copyright 1980,
National Academy of Sciences.

The other potential area of significance for excessive Fe/Zn ratios is in the formulation of vitamin–mineral supplements. We found that, in the 1977 Physician's Desk Reference (55), of 21 such products specifying both zinc and iron content, only 3 had a Fe/Zn ratio less than 3.0 (50). In the 1981 PDR (56), of 18 such products, 6 had a Fe/Zn ratio of less than 3.0 (Table IV).

Tin:Zinc Interaction

The rodent experiments of de Groot and co-workers (57,58) demonstrated a reversal of the hematological effects of high levels of dietary tin by copper and iron. Growth retardation was not reversed by these minerals. Yamaguchi et al. (59) found a significant reduction in serum alkaline phosphatase and lactic dehydrogenase within 90 days of administration of 6 mg Sn/kg body weight as stannous chloride to rats. Because zinc is related to growth, and since alkaline phosphatase and lactic dehydrogenase are zinc-dependent enzymes, Greger and Johnson (60) suspected that dietary tin might adversely affect zinc nutriture. Two groups of rats were fed identical diets, except that the tin content of the control diet was 1 μg/g, and that of the tin-supplemented group, 206 μg. Tin supplementation was provided as stannous chloride. Fecal zinc was 19% higher in the tin-supplemented animals (p <0.05), and zinc concentrations in tibia (p <0.005) and kidney (p <0.05), but not liver, were significantly reduced with tin supplementation.

Studies in Dr. Greger's laboratory also have evaluated the influence of tin or zinc retention in humans (61). Eight healthy male volunteers received their accustomed caloric intakes, 3150 to 3700 kcal, as controlled, metabolic diets of real foodstuffs for two consecutive 20-day cycles; two balance periods –– comprising days 7–12 and days 13–18 of each cycle –– were undertaken. The diets contained an average of 13.5 mg of zinc daily. In a randomized fashion, the rations during one of the dietary cycles had 50 mg of Sn as stannous chloride added. The dietary Sn/Zn

Table IV.   Iron:Zinc Ratio in Propietary Vitamin-Mineral
            Preparations

Brand-name	Manufacturer	Iron (mg)	Zinc (mg)	Fe/Zn ratio
Centrum	(Lederle)	27.0	22.5	1.2:1*
Cluvisol	(Ayerst)	15.0	0.6	25:1
Cluvisol-130	(Ayerst)	15.0	0.6	25:1
Geriplex	(Parke-Davis)	30.0	2.0	15:1
Geriplex-FS	(Parke-Davis)	30.0	2.0	15:1
Megadose	(Arco)	10.0	25.0	0.4:1*
MiCebrin	(Dista)	15.0	1.5	10:1
MiCebrin T	(Dista)	15.0	1.5	10:1
Mayadec	(Parke-Davis)	20.0	20.0	1:1*
Nuclomin	(Miller)	10.0	7.5	1.3:1*
Optilets-500	(Abbott)	20.0	.1.5	13.3:1
Optilets-M-500	(Abbott)	20.0	1.5	13.3:1
Ragus	(Miller)	20.0	2.0	10:1
Sclerex	(Miller)	10.0	3.0	3.3:1
Superplenamins	(Rexall)	24.0	15.0	1.6:1*
Theragran-M	(Squibb)	12.0	1.5	8:1
Theragran-Z	(Squibb)	12.0	22.5	0.5:1*
Vicon with iron	(Glaxo)	30.0	18.5	1.6:1

*denotes a Fe/Zn ratio of 2:1

ratio was 3.7. The mean apparent absorption of zinc on the unsup-
plemented diet was -0.14 mg and -2.1 mg on the tin-supplemented
regimen. These were statistically different (p <0.01), suggesting
that tin influenced intraintestinal zinc metabolism.

As metabolic balance techniques cannot separate the contribution of malabsorption from that of endogenous loss, collaborators
at the Division of Nutrition, University of Sao Paolo Medical
School at Ribeirao Preto, Brazil and I undertook to assess Sn:Zn
interaction using the change-in-plasma-zinc approach. Healthy
volunteers received 12.5 mg of zinc as 55 mg of zinc sulfate in
100 ml of CocaCola either alone (control) or with 25, 50 or 100 mg
of tin as stannous chloride to constitute 2:1, 4:1 and 8:1 Sn/Fe
ratios. We measured the change in plasma zinc concentration at
hourly intervals over a 4-h period. None of the treatments produced any significant decrement in the uptake of zinc in plasma
(61) (Table V). Thus, unlike the dramatic effect of even lesser
ratios of Fe/Zn (above), the plasma appearance of zinc was unaffected by soluble, inorganic, divalent tin ions.

There are several potential resolutions of the apparent dis-
agreement between the metabolic balance data (62) and that seen
with plasma appearance (61). First, the setting of a mixed diet
might have provided additional factors necessary for the expression of the Sn:Fe interaction, or the meal-stimulation of

Table V. Effect of Tin on the Plasma Appearance of
Zinc in Healthy Subjects

Treatment	Change in plasma zinc concentration, µg/dl			
	1-h	2-h	3-h	4-h
12.5 mg of Zn++ a,b	45 + 24d	55 + 11	41 + 9	24 + 7
12.5 mg of Zn++ plus 25 mg of Sn++ c	23 + 10	47 + 19	38 + 20	18 + 15
12.5 mg of Zn++ plus 50 mg of Sn++	27 + 13	43 + 7	38 + 8	28 + 7

a) ingested in 100 ml of CocaCola, mean of 6 individuals
b) zinc as zinc sulfate
c) tin as stannous chloride
d) mean + SD

pancreatic secretion might have played a cooperative role. Our
experiments were conducted with aqueous solutions; carbonate and
phosphate ions were the only potential participating ligands, and
there was negligible pancreatic stimulation. Another possibility
is that tin effects fecal zinc balance not as an inhibitor of
dietary zinc absorption, but as a promoter of endogenous zinc excretion. Calloway and McMullen (63) have shown tin to be essentially quantitatively excreted in the feces. The mechanism of
Sn:Fe interaction merits further investigation.

Implications for Human Nutrition

The tin content of foods has been reported by several investigators (64-67). Storage of food in tin-alloyed cans increases the
tin concentration in foodstuffs; bottled foods have essentially no
tin (67). The practice of coating cans with lacquer -- used widely
in industrialized countries, but not in some developing nations
(65) -- reduces the leaching of tin, but even with lacquer cans,
imperfections in the coating (pinholes) allow contact of foods
with metal. The concentration of tin in selected foods packaged
in various types of cans is illustrated in Table VI.

No official safety standards for daily tin intake have been
established. In 1940, the average tin intake by an adult male in
the U.S. was estimated to be 17 mg/day (68). Children consuming
half a liter of daily of canned, evaporated milk manufactured under prevailing conditions in Ghana (65) would take in between 20
and 50 mg of tin, depending on the storage time of the product.
Volunteers consuming stored army rations ingested 163 mg of tin
daily (63). Greger and Baier(67) have estimated that individuals
who consume 500 g of canned foods stored under refrigeration in
open, unlacquered cans can easily ingest 200 mg of tin daily,
four times the level shown to reduce zinc retention (61). Canning
industries in some developing countries are at a rudimentary state

Table VI. Tin Content of Assorted Canned Foodstuffs

Commodity	Tin μg/g	Type of can	Reference
Sauerkraut	0.6	lacquered	(67)
Tomato juice	1.2	lacquered	(67)
Cranberry sauce	1.4	lacquered	(67)
Portuguese sardines	1.6	unlacquered	(64)
Stewed tomatoes	2.8	lacquered	(67)
Whole beets	3.4	lacquered	(64)
Dietetic carrots	6.4	lacquered	(64)
Condensed tomato soup	7.4	lacquered	(64)
Dietetic spinach	11.8	lacquered	(64)
Moroccan sardines	18.6	unlacquered	(67)
Grapefruit juice	40.9	partially-lacquered	(67)
Evaporated milk*	41.5	unlacquered	(65)
Apple sauce	50.7	partially-lacquered	(67)
Orange juice	53.2	partially-lacquered	(67)
Norwegian herring	66.4	unlacquered	(64)
Crushed pineapple	89.4	unlacquered	(67)
Grapefruit sections	96.4	partially-lacquered	(67)
Portuguese anchovies	129.3	unlacquered	(64)
Tomato sauce	149.8	partially-lacquered	(67)

* stored for 2 months

of technological development, and may provide foods with high tin content. The consequences for zinc nutriture could prove to be a legitimate public health concern.

Magnesium:Zinc Interaction

Since magnesium ion is as divalent cation, magnesium:zinc competition might well exist in the intestine. Magee and Fu (69) showed that a massive enrichment of rat rations, to 0.75% zinc, increased fecal magnesium excretion from 29 to 44%. Geders et al. (70) studied magnesium:zinc interactions in 10 healthy adult volunteers by giving oral doses of 25 mg of zinc as zinc sulfate along with 25, 50 and 75 mg of magnesium as magnesium sulfate, constituting Mg/Zn ratios of 1:1, 2:1 and 3:1. The change in plasma zinc was used as the index of zinc absorption. Mean cumulative increase in the plasma zinc concentration over a 3-h period -- 165, 147 and 149% above baseline levels, respectively -- was unaffected by the dosage of magnesium; thus, at least up to a 3:1 ratio, no inhibitory effect on the intestinal zinc uptake was evident.

Implications for Human Nutrition. The RDA for magnesium in adults is 350 mg for males and 300 mg for females (1). Thus, diets containing the RDAs for both minerals would contain Mg/Zn ratios of 23.3 and 20.0, respectively. Dietary and truly supraphysiological

Mg/Zn ratios must be studied before the possible impact of magnesium on zinc absorption can be assessed.

Conclusion

Competitive mineral–mineral interactions involving chemically similar elements are widespread in nature. Interactions between zinc and iron, and zinc and tin reduce zinc absorption in humans, and pose potential nutritional consequences. Dietary zinc levels may also be of importance in human copper absorption. Natural foods, alone or in combination, tend to have a balance among minerals, but in the formulation of proprietary vitamin–mineral supplements or infant foods, there is the potential for creating nutritionally-significant imbalances. Little is known about the role of other food components in conditioning these interactions. This question should be a priority for future research.

Literature Cited
1. "Recommended Dietary Allowances," 9th Edition; Washington, D.C., National Academy of Sciences, 1980.
2. Hambidge, K.M.: Casey, C.E. In "Gastrointestinal Development and Infant Nutrition," Lebenthal, E., Ed.; Raven Press, New York, 1981, pp. 595-610.
3. Solomons, N.W. In "Nutrition in the 1980s -- Constraints of our Knowledge," White, P.; Selvey, N., Eds.; Alan R. Liss, New York, 1981, pp. 97-127.
4. Solomons, N.W. In "Nutrition Updates," Weininger, J.; Briggs, G.M., Eds., Wiley, New York, in press.
5. "Guidelines for essential trace element preparations for use"; JAMA. 1979, 241, 2051.
6. Hill, C.L.; Matrone, G. Fed. Proc. 1970, 29, 1474.
7. Cousins, R.J. Am. J. Clin. Nutr. 1979, 32, 339.
8. Cousins, R.J. Nutr. Rev. 1979, 37, 97.
9. Bremner, I.; Young, B.W. Biochem. J. 1976, 157, 517.
10. Richards, M.P.; Cousins, R.J. Proc. Soc. Exp. Biol. Med. 1977, 156, 505.
11. Evans, G.W. Proc. Soc. Exp. Biol. Med. 1976, 151, 775.
12. Smith, K.T.; Cousins, R.J.; Silbon, B.L.; Failla, M.L. J. Nutr. 1980, 108, 1849.
13. Smith, K.T.; Cousins, R.J. J. Nutr. 1980, 110, 316.
14. Smith, S.E.; Larson, E.J. J. Biol. Chem. 1946, 163, 29.
15. Duncan, G.D.; Gray, L.F.; Daniel, L.J. Proc. Soc. Exp. Biol. Med. 1953, 83, 625.
16. Grant-Frost, D.R.; Underwood, E.J. Aust. J. Exp. Biol. Med. Sci. 1958, 36, 339.
17. Magee, A.C.; Matrone, G. J. Nutr. 1960, 72, 233.
18. Van Reen, R. Arch. Biochem. Biophys. 1953, 46, 337.
19. Magee, A.C.; Grainger, F.P. Nutr. Rep. Intl. 1979, 20, 771.
20. Spivey Fox, M.R.; Hamilton, R.P.; Jones, A.D.L.; Fry, B.E., Jr.; Jacobs, R.M.; Jones, J.W. Fed. Proc. 1978, 37, 324.

21. Nordstrom, J.W.; Kohrs, M.B.; Howser, M.; Dowdy, R.P. Fed.
    Proc. 1978, 37, 894.
22. Schlicker, S.A.; Cox, D.H. J. Nutr. 1968, 95, 287.
23. Ketcheson, M.R.; Barron, G.P.; Cox, D.H. J. Nutr. 1969, 98,
    303.
24. Cox, D.H.; Schlicker, S.A.; Chu, R.C. J. Nutr. 1969, 98, 459.
25. Chu, R.C.; Cox, D.H. Nutr. Rep. Intl. 1970, 2, 179.
26. Chu, R.C.; Cox, D.H. Nutr. Rep. Intl. 1972, 5, 61.
27. Van Campen, D.R. J. Nutr. 1966, 88, 125.
28. Van Campen, D.R.; Scaife, P.U. J. Nutr. 1967, 91, 473.
29. Evans, G.W.; Grace, C.I.; Hahn, C. Bioinorg. Chem. 1974, 3,
    115.
30. Ogiso, T.; Morujama, K.; Sasaki, S.; Ishimura, Y.; Minato, A.
    Chem. Pharm. Bull. 1974, 22, 55.
31. Van Campen, D.R. J. Nutr. 1967, 97, 104.
32. Evans, G.W.; Grace, C.I.; Votava, H.J. Am. J. Physiol. 1975,
    228, 501.
33. Evans, G.W.; Majors, P.F.; Cornatzer, W.E. Biochem. Biophys.
    Res. Comm. 1970, 40, 1142.
34. Ogiso, T.; Ogawa, N.; Miura, T. Chem. Pharm. Bull. 1979, 27,
    515.
35. Hall, A.C.; Young, B.W.; Bremner, I. J. Inorg. Biochem. 1979,
    11, 57.
36. Fischer, P.W.F.; Giroux, A.; L'Abbe, M.R. Am. J. Clin. Nutr.
    1981, 34, 1670.
37. Greger, J.L.; Zahic, S.C.; Abernathy, R.P.; Bennett, O.A.;
    Huffman, J. J. Nutr. 1978, 108, 1449.
38. Burke, D.M.; DeMicco, F.J.; Taper, L.J.; Ritchey, S.J. J.
    Gerontol. 1981, 36, 558.
39. Taper, L.J.; Hinners, M.L.; Ritchey, S.J. Am. J. Clin. Nutr.
    1980, 33, 1077.
40. Pfeiffer, C.C.; Jenny, E.H. Fed. Proc. 1978, 37, 324.
41. Prasad, A.S.; Brewer, G.J.; Schoomaker, E.B.; Rabbini, P.
    JAMA 1978, 240, 2166.
42. Cox, D.H.; Harris, D.L. J. Nutr. 1960, 70, 514.
43. Magee, A.C.; Fu, S.C. Nutr. Rep. Intl. 1979, 19, 649.
44. Pollack, S.; George, J.N.; Reba, R.C.; Kaufman, R.M.; Crosby,
    W.A. J. Clin. Invest. 1965, 44, 1470.
45. Forth, W. In "Trace Element Metabolism in Animals" Mills,
    C.F., Ed., Livingston, Edinburgh, 1970, pp. 298-310.
46. Flanagan, P.R.; Haist, J.; Valberg, J.D.; Valberg, L.S. J.
    Nutr. 1980, 110, 1754.
47. Hamilton, D.L.; Bellamy, J.E.C.; Valberg, J.D.; Valberg, L.S.
    Canad. J. Physiol. Pharmacol. 1978, 56, 384.
48. Solomons, N.W.; Jacob, R.A.; Pineda, O.; Viteri, F.E. J. Nutr.
    1979, 109, 1519.
49. Solomons, N.W.; Jacob, R.A.; Pineda, O.; Viteri, F.E. J. Lab.
    Clin. Med. 1979, 94, 335.
50. Solomons, N.w.; Jacob, R.A. Am. J. Clin. Nutr. 1981, 34, 475.

51. Viteri, F.E.; Garcia, R.; Torun, B. XXVII National Medical Congress of Guatemala, Guatemala City, Abstracts 1976, p.56.
52. Viteri, F.E.; Garcia, R.; Torun, B. In "The 1976 Annual Report of the Institute of Nutrition of Central America and Panama," Pan American Health Organization, Guatemala City, 1977, p. 103.
53. Pennington, J.A.T.; Church, H.N. "Bowes and Church's Food Values of Portions Commonly Used," J.B. Lippincott, Philadelphia, 1980.
54. Walravens, P.A.; Hambidge, K.M. Am. J. Clin. Nutr. 1976, 29, 1114.
55. "Physicians' Desk Reference," Medical Economics Co., Oradell, NJ, 1977.
56. "Physicians' Desk Reference," Medical Economics Co., Oradell, NJ, 1977.
57. de Groot, A.P.; Feron, V.J.; Til, H.P. Food Cosmet. Toxicol. 1973, 11, 955.
58. de Groot, A.P. Food Cosmet. Toxicol. 1973, 11, 955.
59. Yamaguchi, M.; Saito, R.; Okada, S. Toxicol. 1980, 16, 267.
60. Greger, J.L.; Johnson, M.A. Food Cosmet. Toxicol. 1981, 19, 163.
61. Johnson, M.A.; Baier, M.J.; Grager, J.L. Am. J. Clin. Nutr. 1982, 35, .
62. Solomons, N.W.; Marchini, J.S.; Duarte Favaro, R.M.; Vannuchi, H.; Dutra de Oliveira, J.E. Submitted for publication.
63. Calloway, D.H.; McMullen, J.J. Am. J. Clin. Nutr. 1966, 18, 1.
64. Schroeder, H.A.; Balassa, J.J.; Tipton, I.H. J. Chron. Dis. 1964, 17, 483.
65. Woolfe, M.L.; Manu-Tawiah, W. Ecol. Food Nutr. 1977, 6, 133.
66. Capar, S.G. J. Food Safety 1978, 1, 241.
67. Greger, J.L.; Baier, M.J. J. Food Sci. 1981, 46, 1751.
68. Kehoe, R.A.; Cholak, J.; Story, R.V. J. Nutr. 1940, 19, 579.
69. Magee, A.C.; Fu, S.C. Nutr. Rep. Intl. 1979, 19, 343.
70. Geders, J.M.; Brown, T.D.; Crews, M.G. Fed. Proc. 1981, 40, 938.

RECEIVED October 27, 1982

INDEX

INDEX

A

Absorption
apparent
definition 69
equation for determining 224
and blood zinc concentration 70–72
comparing stable and radioisotope 54–55
effect of calcium 116, 217
effect of calcium and phytate 151, 152*f*
effect of diet 108–10
effect of fermentation period in
bread 218
effect of fiber 116, 191, 216, 217
effect of iron 116
effect of phytate 116, 189, 190
effect of protein 108, 216, 217
and excretion, compared 93–95
from foods of plant origin 116
fractional, definition 69
GI
definition 68
effect of age 76–79
effect of food 76*t*, 78
effect of sex 78, 79*f*
effect of various minerals 247–69
variability 73–75
homeostatic control 234
kinetic model 65, 66*f*, 67*f*
mechanism 234, 244
and metabolic interactions 62*f*, 63
methodology for measurement ..67–71, 73
net, definition 68
from products made with soy 189, 190
stable isotope from different
sources 56–58
true, definition 69
variability 98
Acute zinc deficiency 85–87
Age effects on GI zinc absorption 76–79
Alkaline phosphatase and zinc
deficiency 6, 8, 11
Ammonia levels in zinc deficiency 7, 10
Analytical considerations for zinc
bioavailability 44–48
Apparent absorption
for determining zinc bioavailability 33, 34*f*
definition 69
equation for determining 224
Ascorbic acid, effect on zinc
utilization 122–24
Atomic abundance, natural, of stable
isotopes of zinc 35*t*

B

Balance studies for determining zinc
bioavailability 32
Balance, zinc
effect of calcium and
phosphorus 224–29
effect of fiber 130–42, 192*t*
effect of oxalic acid 134–42
effect of phytate 163–65
Basolateral membrane, zinc transfer 235
Bioassay of zinc, methodology 198, 206
Bioavailability
analytical considerations 44–48
from cereal-based foods 187–94
in cereals 201–4, 208
determining by apparent
absorption 33, 34*t*
determining by balance studies 32
determining by load tests 33
determining by radioisotopic labels 33
determining by stable isotope
labels 33, 35–37, 41–59
of dietary zinc, determining by
stable isotopes 31–40
effect of calcium 178–79
effect of calcium and phosphorus ..224–29
effect of fiber 159–60
effect of fiber and oxalic acid 129–42
effect of magnesium 180
effect of phytate 145–56, 159–60
effect of phytate and fiber 204–6
effect of phytate/zinc molar
ratio 175–76, 179
food, factors affecting 188*t*
in infant formulas 197–201
mathematical calculations 44–47
methodological issues 48–55
phytate/zinc molar ratio,
effect 159–60
in potential fortification sources 188*t*
from processed soybean
products 175–82
in vegetarian diets 115–25
Biological interactions of chemically
similar minerals 248
Biological labeling methodology 53, 54*t*
Blood zinc concentration and zinc
absorption 70–72
Bread, effect of fermentation period
on zinc absorption 218
Brush border membrane, zinc
uptake 234, 239

C

Calcium
complexation 149, 150f
effect on zinc absorption 116, 151,
152f, 217
effect on zinc bioavailability 178–79
interaction 223
and phosphorus, effect on zinc
balance 224–29
Calcium–phytate–zinc
interrelationship 147
Cereal-based foods
zinc bioavailability 187–94
zinc content 186t
Cereals, zinc bioavailability 201–4, 208
Chemical relationships of phytate 149
Chronic zinc deficiency 88–90
Citrate, effect of zinc on uptake 240
Clinical syndrome and zinc
deficiency 83–93
Compartmental model
of zinc metabolism 95–97
for zinc absorption 65–67
Complexation
effect of processing 173
phytate and calcium 149, 150f
phytate and zinc 149, 150f
phytate, calcium, and zinc 149, 150f
Connective tissue, effect of zinc
deficiency 7, 11
Content in vegetarian diets 115–16
Content, iron, of various foods 264t
Contraceptives, oral, effect on
zinc level 110–11
Copper–zinc interactions 253–56

D

Dairy products, as source of
dietary zinc 23, 24t
Deficiency
acute 85–87
and alkaline phosphates 6, 8, 11
in ammonia levels 7, 10
chronic 88–90
and clinical syndromes 83–93
and deoxythymidine kinase 11
effects on gonadal functions 7, 12
and erythrocyte zinc level 6, 8, 11
and hormones 8, 9
and hypercatabolism of fat 11
and hypogonadism 1
induced, experimental method 2–4
and lactic dehydrogenase 6, 8, 11
and leukocyte zinc level 6, 8, 11
and nitrogen excretion 6
overview 1–13

Deficiency—*Continued*
and plasma zinc level 6, 8, 11
and protein synthesis 11
and ribonuclease 6, 8, 11
and sperm count 8–9, 12
subacute 90–93
and testicular function 1
and testosterone 9
and urinary excretion 6
variability 151t
and weight loss 5
Deoxythymidine kinase and zinc
deficiency 11
Diet, effect on zinc absorption 108–10
Dietary zinc
bioavailability assessment 31–40
in food(s) 18, 21, 22t
variation by year 19t
Diets, U.S., zinc levels 16t
Distribution 33, 34f
Distribution of isotope in body 63, 64f

E

Elimination 33, 34f
Endogenous zinc secretion into
intestines 236
Enzymes, levels in plasma in zinc
deficiency 6, 8, 11
Erythrocyte zinc level and zinc
deficiency 6, 8, 11
Excretion, nitrogen and zinc
deficiency 6
Excretion, zinc 33, 34f
Excretion
of labeled zinc 51, 52f
and zinc absorption, compared 93–95
effect of fiber 132–42
urinary, and zinc deficiency 6
Extrinsic factors 116

F

Fat, hypercatabolism and zinc
deficiency.................................... 11
Fermentation period in bread, effect
on zinc absorption 218
Fiber
effect on zinc absorption 116, 191,
216, 217
effect on zinc balance 130–42, 192t
effect on zinc bioavailability 129–42,
159–60
effect on zinc excretion 130–42
effect on zinc utilization 121–22
and phytate, effect on
bioavailability 204–6
Fish, as source of dietary zinc 18, 21, 22t

Food(s)—*See also* Cereal-based foods, Dairy products, Diets, Grain products, Meat, Poultry, Soy products
effects on GI zinc absorption 76t, 78
iron content 264t
minor source(s) of dietary zinc 25, 27t
tin content 268t
Food and nutrition policy in Norway .. 211
Food products, phytate/zinc molar ratio 191
Food supply, U.S.
evaluation methods 17–18
evaluation results 18–27
zinc levels 15–29
Fortification of foods with zinc 186–88
Fractional zinc absorption 69

G

Gas chromatograph mass spectrometry (GC/MS) 36–37
GI absorption
definition 68
effect of various minerals 247–69
mechanism 234
variability 73–75
Glutathione, effect of zinc on uptake .. 240
Gonadal functions, effect of zinc deficiency 7, 12
Grain products, as source of dietary zinc 25, 26t
Growth, chick, effect of phytate 146f
Growth, rats, effect of phytate 148f

H

Homeostatic control of zinc absorption 234
Homeostatic fluxes 153
Hormones and zinc deficiency 8, 9
Humans
determining zinc bioavailability, balance studies 32
kinetic model for zinc 65–67
RDA 15, 28–29
steady state zinc parameters 65t
zinc absorption 38–39
zinc bioavailability
determining by apparent zinc absorption 33, 34t
determining by radioisotopic labels 33
determining by stable isotope labels 33, 35–37, 41–59
determining by zinc load tests 33
zinc deficiency overview 1–13

Hypercatolism of fat and zinc deficiency 11
Hypogonadism and zinc deficiency 1

I

Induced zinc deficiency, experimental method 2–4
Infant formulas, zinc bioavailability 197–201
Interactions
biological, of chemically similar minerals 248
calcium–zinc 223
copper–zinc 253–56
iron–zinc 194, 257–64
magnesium–zinc 268
Intracellular interaction of zinc with various compounds 235
Intracellular interactions, mineral–mineral 251
Intraluminal interactions, mineral–mineral 251
Intrinsic factors 116
Iron
content of various foods 264t
effect on zinc absorption 116
recommended dietary allowances 265t
Iron–zinc interaction 194, 257–64
Iron–zinc ratio in vitamin–mineral preparations 266t
Isotope, stable—
See also Radioisotope
methodology of analysis 36–37
natural atomic abundance 35t
precision measurements 36t
for research, history 35–36
used for assessment of bioavail-ability of zinc 31–40, 41–59
Isotopic constitution of zinc 42, 43f
Isotopic measurement, methodology .. 48–50

K

Kinetic model of zinc metabolism 65–67, 95–97

L

Lactic dehydrogenase and zinc deficiency 6, 8, 11
Leukocyte zinc level and zinc deficiency 6, 8, 11
Levels in U.S. diets 16t
Levels in U.S. food supply 15–29
Load tests for determining zinc bioavailability 33
Luminal surface, zinc uptake 237

M

Magnesium, effect on zinc
    bioavailability 180
Magnesium–zinc interactions 268
Mathematical calculation, zinc
    bioavailability 44–47
Measurement uncertainty 46–47, 50t
Meat, as source of dietary zinc 18, 21, 22t
Mechanism of zinc absorption 234, 244
Metabolic balance—See Balance
Metabolic interactions and zinc
    absorption 62f, 63
Metabolism, compartmental model 95–97
Metallothionein 235
Methodological issues in zinc
    bioavailability 48–55
Methodology
    balance study 108, 160–62, 169
    of biological labeling 53, 54t
    comparing stable and radioisotope 53, 54t
    of isotopic measurement 48–50
    mineral balances determination 128–37
    neutron activation analysis 48
    radioisotope uptake 213
    slope ratio technique 174–75
    of stable isotope analysis 36–37
    for zinc absorption
        measurement 68–71, 73, 213,
                                    236–37
    zinc bioassay 198, 206
    zinc evaluation in U.S. food
        supply 17–18
Micronutrients, effect on zinc
    utilization 193
Minerals—See also Iron, Tin,
    Magnesium, Copper
    effect of GI absorption 247–69
Mineral–mineral interactions,
    intracellular 251
    intraluminal 251
    mucosal 251
    serosal 253
Molar ratio, phytate/zinc
    effect on zinc bioavailability 159–60
    in food products 191
    soybean products, effect on zinc
        bioavailability 175–76, 179
Mucosal interactions, mineral–mineral 251
Myoinositol hexakis(dihydrogen
    phosphate)—See Phytate

N

Net absorption, definition 68
Neutron activation analysis (NAA) 36–37
    methodology 48
Nitrogen excretion and zinc deficiency   6

Norway, nutrition and food policy 211
Nutrient density 28
Nutritional status of vegetarians 117–19

O

Occurrence in nature 31
Omnivores
    zinc utilization compared to
        vegetarians 119–21
    effect of ascorbic acid on zinc
        utilization 122–24
    effect of fiber on zinc utilization ..121–22
Oral contraceptives, effect on zinc
    level 110–11
Overview of zinc deficiency 1–13
Oxalic acid
    effect on zinc balance 134–42

P

Phytate
    biology and chemistry 145–47
    chemical relationships 149
    complexation 149, 150f
    content in Middle Eastern breads .. 155t
    effect on chick growth 146f
    effect on rat growth 148f
    effect on serum zinc 165, 166t, 170t
    effect on zinc absorption ..116, 151, 152f
    effect on zinc balance 163–65
    effect on zinc bioavail-
        ability 154–56, 159–60, 187
    and fiber, effect on bioavail-
        ability 204–6
Phytate–calcium–zinc inter-
    relationship 147
Phytate/zinc molar ratio
    effect on zinc bioavailability 159–60
    in food products 191
    soybean products, effect on zinc
        bioavailability 175–76, 179
2-Picolinate, effect of zinc on uptake .. 240
Plasma levels of enzymes in zinc
    deficiency 6, 8, 11
Plasma levels of zinc in zinc
    deficiency 6, 8, 11
Poultry, as source of dietary
    zinc 18, 21, 22t
Pregnancy, effect on zinc level 110–11
Processing, effect on complexation 173
Protein, effect on zinc
    absorption 108, 191, 193, 216, 217
Protein synthesis and zinc deficiency .. 11

R

Radioisotope and stable methodology
    compared 53, 54t

Radioisotope and stable zinc
absorption compared 54–55
Radioisotope uptake studies 33, 224,
237–38
methodology 213
Recommended dietary allowances
(RDA)
for iron 265t
for zinc 15, 28–29, 265t
Ribonuclease and zinc deficiency 6, 8, 11

S

Secretion, endogenous, zinc into
intestines 236
Serosal interactions, mineral–mineral 253
Serum zinc, effect of phytate 165, 166t, 170t
Sex and GI zinc absorption 78, 79f
Smell dysfunction 95, 98–101
Soy products, absorption 189, 190
Soybean products
phytate/zinc molar ratio, effect on
zinc bioavailability 175–76, 179
processed, zinc bioavailability 175–82
Sperm count and zinc deficiency 8–9, 12
Stable isotope(s)—*See also* Isotope,
stable
for determining zinc bioavail-
ability 33, 35–37, 41–59
Stable and radioisotope methodology
compared 53, 54t
Stable and radioisotope zinc
absorption compared 54–55
Steady state zinc parameters in
humans 65t
Subacute zinc deficiency 90–93

T

Taste dysfunction 95, 98–101
Tin content of various foods 268t
Testicular function and zinc deficiency 1
Testosterone and zinc deficiency 9
Thermal ionization mass spectrometry
(TI/MS) 36–38
Tissue, zinc concentration 101t
Tin–zinc interactions 265–267
Transfer across basolateral
membrane 235
Transfer, mineral 249
Transport by vascularly perfused rat
intestine 233–44

Transport, mineral 249
True zinc absorption, definition 69

U

Uptake
of citrate, effect of zinc 240
of glutathione, effect of zinc 240
of isotope in body 63, 64f
methodology, radioisotope 213
mineral 249
of 2-picolinate, effect of zinc 240
studies, radioisotope 240
of zinc at brush border
membrane 234, 239
of zinc at luminal surface 237
Urinary excretion and zinc deficiency 6
Utilization
effect of ascorbic acid 122–24
effect of fiber 121–22
effect of micronutrients 193
in growing children 111
vegetarians and omnivores
compared 119–24

V

Variability of zinc deficiency 151t
Vascularly perfused rat intestine,
transport 233–44
Vegetarians
ascorbic acid effect on zinc
utilization 122–24
effect of fiber on zinc utilization 121–22
nutritional status 117–19
zinc utilization compared to
omnivores 119–21
Vegetarian diets
zinc bioavailability 115–25
zinc content 115–16
Vitamin–mineral preparations, iron/
zinc ratio 266t

W

Weight loss and zinc deficiency 5

Z

Zinc, recommended dietary
allowances 15, 28–29, 265t
Zinc–phytate–calcium interrelation-
ship 147

Indexing by Janet Dodd
Production by Robin Giroux and Anne Bigler

Elements typeset by Service Composition Co., Baltimore, MD
Printed and bound by Maple Press, Co., York, PA